Caring for Creation

Caring for Creation

Hope in Difficult Times

Daniel E. Lee

LEXINGTON BOOKS
Lanham • Boulder • New York • London

Published by Lexington Books
An imprint of The Rowman & Littlefield Publishing Group, Inc.
4501 Forbes Boulevard, Suite 200, Lanham, Maryland 20706
www.rowman.com

6 Tinworth Street, London SE11 5AL

British Library Cataloguing in Publication Information Available

Library of Congress Control Number: 2019949602

Contents

Acknowledgments

Part of the work on this volume was done while I was on sabbatical from teaching. I am indebted to Augustana College for providing this sabbatical support. I am also indebted to the students during more than four decades of teaching at Augustana College who have asked insightful questions that have stimulated my thinking. I am particularly appreciative of the students who were enrolled in a special topics course on individual rights and the common good which I team-taught with Steven C. Bahls, president of Augustana College. They are Ebony Allen, Imam Saad Baig, Grant Bell, Alex Bruozis, Andrew Ellison, Megan Funke, Bo Goshorn, Annie Hildner, Samantha Kammerman, Sarah Moore, Vanessa Reyes, and Micaela Teronnez. In this seminar I presented some of the ideas that came to fruition in this book.

Prologue

In *Walden,* Henry David Thoreau (1817–1862) observes, "Heaven is under our feet as well as over our heads."[1] As Thoreau knew so well, each season brings its own set of joys when we allow ourselves to be immersed in nature. The music of trickling streams of water from melting snow as lush green new blades of grass poke through the reawakening ground that has shed its winter coat. The cool of a summer evening providing respite from the heat of day as fireflies punctuate the darkness with pin-point bursts of light. The brilliantly clad scarlet maple trees of autumn, their bright red leaves contrasting sharply with the still-green grass, a powerful display of complementary colors one does not have to be an artist to appreciate. The late afternoon winter sun casting pastel blue shadows on fields blanketed with snow with the sky near the horizon touched lightly with bluish-green the color of a robin's egg.

Sharing this wonderful world with us are creatures both large and small. Migrating warblers of many different colors add their music to the melodious orchestra of the natural world as they flit about trees just beginning to leaf out. Black-masked raccoons quietly look for their next meal as a full moon bathes the landscape in soft, soothing light. V-shaped formations of Canada geese, communicating in a language only they understand, head south as the crisp days of autumn signal the coming of winter. Bald eagles with their immense seven-foot wing spans quietly soar above open water in the cold of winter, their keen eyes on the lookout for fish that have injudiciously strayed too close to the surface of the water.

In *Walden*, Thoreau further observes, "We can never have enough of nature. We must be refreshed by the sight of inexhaustible vigor, vast and titanic in its features . . , the wilderness with its living and decaying trees, the thunder-cloud, and the rain which lasts three weeks and produces freshets."[2] He is right about that. As those who allow themselves to be immersed in nature know, there is something grand and glorious and good about nature, whether experienced in the massive panorama of wilderness areas or in microcosm in the joy of the flowering of a bright red amaryllis in a pot on a sunny windowsill.

In *A Sand County Almanac*, published nearly a century after *Walden*, conservationist Aldo Leopold (1887–1948), using words both simple and eloquent, gives expression to his love for the world of nature as he recounts his experiences at the worn-out farm he bought, a farm which became his family's weekend refuge from "too much modernity."[3]

Leopold observes with respect to May:

> When dandelions have set the mark of May on Wisconsin pastures, it is time to listen for the final proof of spring. Sit down on a tussock, cock your ears at the sky, dial out the bedlam of meadowlarks and redwings, and soon you may hear it: the flight-song of the upland plover, just now back from the Argentine . . . .
>
> If your eyes are strong, you may search the sky and see him, wings aquiver, circling among the wooly clouds.[4]

August is a time to rejoice in the artistry of nature. Leopold shares a treasured secret:

> I know a painting so evanescent that it is seldom viewed at all, except by some wandering deer. It is a river who wields the brush . . . . The work begins with a broad ribbon of silt brushed thinly on the sand of a receding shore. As this dries slowly in the sun, goldfinches bathe in its pools, and deer, herons, killdeers, raccoons, and turtles cover it with a lacework of tracks . . . . [W]hen I see the silt ribbon turning green with Eleocharis, I watch closely thereafter, for this is the sign that the river is in a painting mood . . . .[5]

September is a time for music:

> There is a peculiar virtue in the music of elusive birds. . . . What one remembers is the invisible hermit thrush pouring silver chords from impenetrable shadows; the soaring crane trumpeting from behind a cloud; the prairie chicken booming from the mists of nowhere; the quail's Ave Maria in the hush of dawn. No naturalist has even seen the choral act, for the covey is still on its invisible roost in the grass, and any attempt to approach automatically induces silence.[6]

Thoreau and Leopold both exemplify love of the world of nature and the joys that accompany immersing oneself in it and savoring its beauty. Unfortunately, in the history of the American experience, they are the exception, rather than the rule.

EXPLOITATION AND DESTRUCTION

The author of I Timothy in the Christian New Testament observed, "For the love of money is a root of all kinds of evil . . ." (I Timothy 6.10 NRSV). And indeed, love of money has resulted in the exploitation and destruction of much of the North American continent, as well as Australia, New Zealand, and other parts of the world where Europeans saw only economic opportunity and failed to appreciate the intrinsic beauty and goodness of the land.

The American experience is the story of the annihilation of passenger pigeons, beautiful birds which once numbered in the billions but which were slaughtered to the point of extinction. Potawatomi tribal leader Simon Pokagon, seeing what was happening, wondered what type of divine punishment would be "awaiting our white neighbors who have so wantonly butchered and driven from our forests these wild pigeons, the most beautiful flowers of the animal creation of North America."[7]

The American experience is the story of the massive slaughter of bison, which had once numbered somewhere between 30 and 50 million—a horrendous slaughter that drove them to the brink of extinction.[8] Bison were the source of sustenance and just about everything else needed by the Native American tribes living on the Great Plains in what is now the United States and on the prairies of what is now Canada—tribes that included the Blackfoot, Cheyenne, Comanche, Crow, Lakota, and many others. The slaughter of bison by white hunters and others who had no appreciation of Native American culture threatened the very existence of these tribes.

The American experience is the story of the Love Canal on the eastern edge of Niagara Falls, New York, which an article in a publication of the U.S. Environmental Protection Agency has labeled "one of the most appalling environmental tragedies in American history."[9] In a news story entitled "Upstate Waste Site May Endanger Lives" that ran on the front page of the *New York Times*, Donald G. McNeil, Jr. reported, "Twenty five years after the Hooker Chemical Company stopped using the Love Canal here as an industrial dump, 82 different compounds, 11 of them suspected carcinogens, have been percolating upward through the soil, their drum containers rotting and leaching their contents into the backyards and basements of 100 homes and a public school built on the banks of the canal."[10]

The American experience is the story of the Berkeley Pit, a huge open-pit copper mine in Butte, Montana, which, with mining operations having been discontinued in 1982, is filling with contaminated water.[11] The water, which is very acidic with high concentrations of arsenic, copper, cadmium, cobalt, iron, manganese, zinc, and sulfate, is lethal for geese and other waterfowl who make the terrible mistake of landing on what they believe to be a natural lake.[12]

The American experience is the story of excessive use of artificial fertilizers—excessive use that pollutes streams, rivers, and lakes. Pollution that contributes to eutrophication—the rampant growth of algae which ultimately results in oxygen levels so low that fish and other forms of aquatic life can no longer survive.[13] Today, as a result of eutrophication, there are no fish to be found in significant portions of the northern part of the Gulf of Mexico.[14]

Unfortunately, these are but a few examples of what happens when expediency and love of money take priority over appreciating and caring for nature. A prophecy believed to be Native American in origin observes, "When the last tree is cut down, the last fish eaten, and the last stream poisoned, you will realize that you cannot eat money."[15] Such is the legacy of exploitation of land driven by love of money.

STRUGGLING TO SURVIVE AND THE ECONOMIC TETHER

Those of us who teach like neat categories—categories that simplify complex matters in ways that make them easier for students to understand. And so it is tempting to try to reduce environmental matters to "love of the world of nature" versus "love of money."

There is, of course, a good deal of truth that underlies this dichotomy, examples of which have been noted above. Yet this dichotomy oversimplifies matters in ways that can be misleading. Take, for example, an anonymous poem published in 1692 which tells of clearing the land "with Ax, with Might and Strength" by cutting down "trees so thick and strong" which "with Fire, most furiously to ashes did confound."[16]

At first glance, this might appear to be a poem about deforestation driven by love of money. And indeed, history is replete with examples of timber companies owned by wealthy individuals motivated by greed—companies that cut down verdant stands of virgin timber, leaving environmental devastation in their wake.

There is, however, another way to interpret this poem. It might be about pioneers struggling to survive. Pioneers who cleared the land to open up space for planting corn, wheat, and other crops necessary for their survival. In short, there is a huge difference between deforestation fueled by greed, on the one hand, and, on the other hand, deforestation that is part of an economic struggle to survive, even though both result in damage to ecosystems.

This is true in other situations as well. In a subsequent chapter, attention will be focused on environmental damage resulting from the use of artificial fertilizers and pesticides. In some cases, this damage is done by corporations and other agricultural businesses whose owners care only about how much money they can make. But in other cases, this environmental damage is done by family farmers struggling to keep their noses above water. Writing in the latter part of the twentieth century, John Fraser Hart, who taught at the University of Minnesota for a number of years, observed, "The cost of growing corn has continued to rise, and the farmer's profit on each bushel of corn, or on each unit of anything else he produces, has been steadily reduced as the gap has narrowed between rising costs and stable or falling prices." He noted that economic realities "forced [farmers] to start using a whole arsenal of agricultural chemicals" in an effort to increase yields.[17]

In *A Sand County Almanac*, Aldo Leopold observes, "It of course goes without saying that economic feasibility limits the tether of what can or cannot be done for land. It always has and it always will."[18] He is right about that. It is very easy to be moralistic about environmental issues while ignoring underlying economic realities. Moralizing, however, accomplishes very little if one does not include economic feasibility in the equation. That is where the economic tether comes into play.

It is for this reason that the Leopold Center for Sustainable Agriculture at Iowa State University in Ames, Iowa (named after Aldo Leopold), states on its website that it was created "to identify and develop new ways to farm profitably while conserving natural resources as well as reducing negative environmental and social impacts."[19] One of the greatest strengths of the observations Leopold makes in *A Sand County Almanac* is that he recognized both the importance of conservation and the economic realities that farmers and others face. Ethics and economics both can go—and should go—hand-in-hand. Indeed, in some situations, such as the slash-and-burn destruction of the rain forests in the Amazon basin by indigenous groups struggling to survive, economic justice might even be a prerequisite for environmental justice.[20]

At the same time, the economic part of the equation can be a double-edged sword. In *Populorum Progressio*, Paul VI observes, "The pursuit of life's necessities is quite legitimate . . . . But the acquisition of worldly goods can lead men to greed, to the unrelenting desire for more . . . ." The result, he warns, can be falling "prey to avarice and soul-stifling materialism."[21] That is the ever-present danger if one focuses only on the economic aspects of agriculture or any other endeavor, without taking stewardship and environmental impact into consideration.

INTRINSIC VALUE

In *A Sand County Almanac,* Leopold proposes an ethic that "changes the role of *Homo sapiens* from conqueror of the land-community to plain member and citizen of it. It implies respect for his fellow-members, and also respect for the community as such."[22] He explains, "The land ethic simply enlarges the boundaries of the community to include soils, waters, plants, and animals, or collectively: the land."[23]

It has long been known, of course, that the world of nature is of instrumental value to humankind. It provides timber which can be cut down and sawed into lumber used to construct houses. It provides soil which can be used to raise wheat and corn, vital commodities for human consumption, as well as for livestock. It provides mineral resources which can be used to make steel, aluminum, and other materials vital for manufacturing automobiles, washing machines, and many of the other things that are important parts of modern life. What Leopold is suggesting, however, is that the world of nature is more than something that is just of instrumental value. He is suggesting that it also has intrinsic value—that is to say, that it is of value in and of itself, rather than simply something to be exploited by humankind.

But why might this be the case? In a book entitled *In Defense of the Land Ethic*, J. Baird Callicott wrestles with this question. One possible answer that he explores is the variation of creation theology found in Genesis 2, which he suggests can provide a basis for arguing that all members of the biotic community possess intrinsic value.[24] While Callicott places the strongest emphasis on the creation story that appears in Genesis 2, which makes reference to Adam being assigned the task of caring for the Garden of Eden (Genesis 2.15), an even stronger statement of the intrinsic goodness of creation can be found in Genesis 1:

> Then God said, "Let the earth put forth vegetation: plants yielding seed, and fruit trees of every kind on earth that bear fruit with the seed in it." And it was so. The earth brought forth vegetation: plants yielding seed of every kind, and trees of every kind bearing fruit with seed in it. And God saw that it was good. . . .
>
> And God said, "Let the waters bring forth swarms of living creatures, and let birds fly above the earth across the dome of the sky." So God created the great sea monsters and every living creature that moves, of every kind, with which the waters swarm, and every winged bird of every kind. And God saw that it was good. . . .
>
> And God said, "Let the earth bring forth living creatures of every kind: cattle and creeping things and wild animals of the earth of every kind." And it was so. God made the wild animals of the earth of every kind, and the cattle of every kind, and everything that creeps upon the ground of every kind. And God saw that it was good. (Genesis 1:11–24 NRSV)

The Genesis 1 creation story is less appealing to Callicott and other environmentalists, however, because the passage from Genesis 1 quoted above is followed by the story of the creation of humankind, who are given "dominion over the fish of the sea, and over the birds of the air, and over the cattle, and over all the wild animals of the earth, and over every creeping thing that creeps upon the earth" (Genesis 1:26 NRSV)—a passage which has often been interpreted as giving humankind license to dominate and exploit nature, a view that has been exceedingly destructive to other species and to the environment. That, however, is not the only interpretation that might be given to this passage. A footnote to this passage in the *New Oxford Annotated Bible* suggests, "Human dominion, corresponding to God's rule, is to be benevolent and peaceful"[25] Nevertheless, the fact that in the course of the centuries exploitation and destruction, rather than benevolence and peacefulness, have often typified humankind's relationship to nature understandably makes Callicott and other environmental philosophers reluctant to reference the creation story in Genesis 1,[26] notwithstanding the strong affirmation of the goodness of creation. Rather, Callicott is more comfortable with the creation story in Genesis 2, about which he observes, "If [Adam] is not, in Aldo Leopold's terms, 'a plain member and citizen' of Eden, neither is he its conqueror or master. Rather, Adam's role is to be custodian or steward of creation."[27]

The other moral theory which Callicott believes offers the greatest promise in this regard is bio-empathy.[28] When we become aware of and experience the richness of the natural world, which Leopold so engagingly describes in *A Sand County Almanac*, we come to experience bio-empathy, which in turn results in our becoming more concerned about and responsive to the needs of other members of the biotic communities of which we are a part. Callicott concludes that "[the creation story found in Genesis 2] and bio-empathy appear to me to provide most effectively for the intrinsic value of other species." He adds, "Each has wide appeal to different and complementary segments of the public, and each is relatively simple and straightforward."[29] His observations are useful contributions to this discussion.

Leopold, it should be noted, is by no means the only environmentalist who views humankind as part of a broader community. For example, in *Laudato Si'*, Francis, drawing inspiration from Saint Francis of Assisi (whose name Francis chose when he was elected pope) and Saint Bonaventure, who "would call creatures, no matter how small, by the name of 'brother' or 'sister,'" observes:

> Such a conviction cannot be written off as naïve romanticism, for it affects the choices which determine our behavior. If we approach nature and the environment without this openness to awe and wonder, if we no longer speak the language of fraternity and beauty in our relationship with the world, our attitude will be that of masters, consumers, ruthless exploiters, unable to set limits on their immediate needs. By contrast, if we feel intimately united with all that exists, then sobriety and care will well up spontaneously.[30]

INSTRUMENTAL VALUE

While it is possible to argue (at least if one uses the Genesis creation stories as reference points) that all species possess *intrinsic* value, it does not follow that all species possess the same *instrumental* value. Take, for example, the *Anopheles* mosquito, which transmits malaria. Many would say that it not only lacks positive instrumental value but, though the term is a bit awkward, might even be said to have negative instrumental value—that is to say, be detrimental to those who contract malaria as a result of being bitten by an *Anopheles* mosquito. The notion of negative instrumental value might also be applied to black legged ticks (*Ixodes scapularis*), which spread Lyme disease, pack rats (*Neotoma cinerea*), which tear apart and make a mess of just about everything they get into, viruses which cause smallpox (*Variola major* and *Variola minor*), spotted knapweed (*Centaurea maculosa*), a non-native plant that has become a noxious weed in fields and pasture land in the western United States, and several other species of plants and animals. Indeed, the negative instrumental value that these and other species have might even be sufficient in magnitude to offset any intrinsic value they might have and, in some instances, justify their eradication. For example, in order to reduce the risk of malaria being transmitted by *Anopheles* mosquitos, many public health authorities recommend spraying with insecticides swamps that are breeding grounds for these mosquitos—a preventive measure many countries have taken.

It might be added that implicit in Leopold's land ethic is a notion of negative instrumental value in some situations—that is to say, some members of the land community might pose a threat to other members of the land community if allowed to run rampant. In *A Sand County Almanac*, this is perhaps clearest in his extolling the virtue of gardening. ("There are two spiritual dangers in not owning a farm. One is the danger of supposing that breakfast comes from the grocery, and the other that heat comes from the furnace. To avoid the first danger, one should plant a garden, preferably where there is no grocer to confuse the issue."[31])

As anyone who has tried her or his hand at gardening knows, raising peas, beans, carrots, corn, and other vegetables is not simply a matter of planting the seeds in springtime and then letting nature run its course, the end result being an abundance of vegetables exactly like those depicted in the colorful seed catalogs which arrive in late winter and lure us into thinking that gardening is easy. If the garden is left unattended, weeds will choke out the peas, the beans, the carrots, the corn, and whatever else was planted, leaving a meager harvest, if any. According to the Genesis 2 creation story, "the Lord God took the man and put him in the Garden of Eden to till it and keep it" (Genesis 2.15 NRSV). Even Adam, it seems, had to do some weeding. If one does not take hoe in hand and chop out the weeds between the rows of corn, beans, and peas in one's garden, they will not grow and flourish.

This leaves us with a difficult question: if we are to respect all members of the biotic communities of which we are a part, does this include the weeds that threaten the peas, beans, carrots, corn, and other vegetables that we have planted in our gardens? Is killing the weeds by chopping them out with a hoe respectful of them? The only plausible answer that might be given to this conundrum is to respond by saying that the negative instrumental value of the weeds—that is to say, the threats that they pose to the food we are raising for human consumption—outweigh whatever intrinsic value plants viewed as weeds might have. The same can be said about mosquitos, black legged ticks, various non-native invasive species, and several other species of plants and animals.

AESTHETIC VALUE

What is true of instrumental value is also true of aesthetic value, though this is even more difficult to quantify. Suffice it to say that not everything that is part of the natural world is perceived to be aesthetically pleasing. Have you ever heard anyone speak in glowing terms of the beauty of pack rats and mosquitoes?

Yet, as noted above, there is much that is beautiful and grand and glorious about nature—beauty that Leopold experienced and to which he gave eloquent expression: "I know a painting so evanescent that it is seldom viewed at all, except by some wandering deer. It is a river who wields the brush . . . ,"[32] and "There is a peculiar virtue in the music of elusive birds . . . ."[33]

In short, when instrumental value and aesthetic value are taken into account, along with intrinsic value, there is a case to be made for saying that all species do not have equal value. At the same time, if one assumes that all species have intrinsic value, it is plausible to argue that, particularly in the case of animals, all species ought to be treated with respect and dignity and not be tortured or otherwise be treated in an inhumane manner.

BONDING WITH NATURE

As Leopold's observations illustrate, there is an experiential dimension to developing a healthy environmental ethic that helps protect the biotic communities of which we are a part while enriching our own lives as well, not in financial terms but in terms of our own growth and development and the quality of life that we experience. Bonding with nature is the key to discovering and realizing these life enriching experiences.

One need not go to national parks or wilderness areas to experience the natural world, though visiting national parks and backpacking in wilderness areas is time well spent. Nor does one need to go to a farm or in other ways spend time in rural areas, though there is much to be gained by so doing. The natural world can be experienced in urban settings, as well as in rural settings. This can be accomplished by spending time in urban parks and other green space that has been preserved. It can be accomplished by going on walks on trails in wooded areas or along the shores of lakes and the banks of rivers. It can even be accomplished by something as simple as putting up a bird feeder and learning to know the rich variety of feathered creatures that come to dine on what is offered at these feeding stations.

Unfortunately, to varying degrees, we all have a tendency to think that we are too busy for birdwatching or hiking trails in wooded areas or along sandy beaches. We have projects to complete, papers to write, products to design, meetings to attend, and all sorts of other things to do which crowd out what might be gained from experiencing the world of nature. The result is that we miss out on the richness of the natural world and are impoverished in the process. Moreover, our detachment and apathy place the natural world at risk, as those motivated by love of money seek to replace green space with housing developments, condominiums, and office buildings, which they believe would make better use of the land. This is one of the greatest tragedies of our time—a tragedy for which all of us, to varying degrees, are responsible.

Two recent studies document this detachment. A 2017 study conducted by the public relations and marketing firm DJ Case and Associates in conjunction with state and federal wildlife and park agencies, determined that adults and children alike spend less time outdoors than did previous generations. Computers, smart phones, and other electronic devices are contributing factors to this growing disconnect from nature.[34] A 2018 study by Lincoln R. Larson *et al.* entitled "Outdoor Time, Screen Time, and Connection with Nature: Troubling Trends among Rural Youth?" reports that while most youth spend time outdoors, they spend far more time indoors with electronic media.[35]

Yet there are some hopeful signs. The DJ Case study reports that many of those interviewed and surveyed indicated that they would like to spend more time outside experiencing nature.[36] And in recent years, recreational visits to national parks have continued to increase. In 2005–2008, the number of recreational visits to national parks averaged 274,413,807 per year. A decade later (2015–2018), the average number of visits per year increased to 321,828,394,[37] a number that probably would have been even greater, were it not the case that many national parks are today suffering from overcrowding with existing facilities pressed to their limits and beyond.

A JOINT VENTURE WITH MOTHER NATURE

Gardening is a superb way of connecting with nature. In the chapter entitled "The Bean Field" in *Walden*, Thoreau observes, "I came to love my rows, my beans. . . . They attached me to the earth, and so I got strength like Antæus."[38]

Gardening is, if you will, a joint venture with Mother Nature—a joint venture that reminds us of the limits of our power and of the importance of humility and gratitude. At the same time, gardening is not something that Mother Nature can handle all by herself. As a joint venture, gardening requires input from those who garden. This includes tilling the soil and preparing it for planting, watering the garden on days when Mother Nature does not provide rain, weeding the garden, and, in the fullness of time, harvesting the vegetables and fruit that the garden produces. As noted, the Genesis 2 creation story states that, "the Lord God took the man and put him in the Garden of Eden to till it and keep it" (Genesis 2.15 NRSV). The same is true of our own Gardens of Eden, be they backyard vegetable gardens or flower beds gracing the front of our homes, for we must till and care for these Gardens of Eden if they are to flourish.

Gardening instills appreciation for the seasons of the year. In northern climates, peas and lettuce can be planted as soon as the soil dries out enough to work up. If the handful of soil that one picks up sticks together in a wet clump, it is too early to work up the ground. If the soil crumbles when squeezed, it is time to get the spade, hoe, and rake out of the tool shed and go to work. Once the soil is worked up, it is time to plant peas, lettuce, radishes, and other cold-weather crops. After planting, one must check the garden every day—sometimes several times a day. In *Mosses from an Old Manse*, Nathaniel Hawthorne (1804–1864) recalls:

> My garden . . . was of precisely the right extent. An hour or two of morning labor was all that it required. But I used to visit and revisit it, a dozen times a day, and stand in deep contemplation over my vegetable progeny, with a love that nobody could share nor conceive of, who had never taken part in the process of creation. It was one of the most bewitching sights in the world, to observe a hill of beans thrusting aside the soil, or a row of early peas, just peeping forth sufficiently to trace a line of delicate green.[39]

When peas and other early crops break through the surface of the soil, marking the rows with bright green newly-formed leaves, the gardening season is officially underway. It is a wonderful time of the year!

Tender crops like beans cannot be planted until all danger of frost has passed. The same is true of sweet corn, which, some say, should not be planted until the leaves of oak trees are the size of squirrels' ears. Squash and cantaloupe are warm weather crops and should not be planted until the soil has warmed up, which doesn't happen until late May or early June in northern climates.

Tomatoes are essential if one is to have a vegetable garden. To minimize bruising during picking and shipping, the tomatoes offered for sale in the typical supermarket have been genetically engineered to be as tough as the bumper of a car, which does not in any way improve their flavor. Tomatoes should not be purchased in a supermarket unless one is desperate for tomatoes. It is far better to purchase tomatoes at a farmers' market or, even better yet, grow them in one's own garden. Early Girl and Big Boy are two varieties of tomatoes that do quite well in many climates.

The greatest joy of gardening, of course, is harvesting the produce. Radishes and lettuce usually come first, with peas not far behind. If one plants an early variety of potatoes such as Norland, it is possible to have new potatoes by early July. (There is an old folk saying which holds that if you plant a row of potatoes on Good Friday, you will have new potatoes by the Fourth of July.) Sweet corn follows shortly thereafter. If sufficient garden space is available, it is good to plant several different varieties of sweet corn that ripen at different times. That will extend the harvest season.

It is widely assumed that when the first frost comes, that is the end of the harvest season. That, however, is not true. Brussels sprouts can be left in the garden until after the first frost, which improves their flavor. Fresh from the garden and steamed or sautéed, they are a tasty addition to a Thanksgiving dinner. Parsnips can be left in the ground all winter and then dug in the spring when the frost has left the ground. As is the case with Brussels sprouts, freezing improves their flavor. Parsnips sautéed in butter are a gourmet's delight—particularly parsnips that were left in the ground all winter (which have much better flavor than the parsnips covered with wax that are sold in supermarkets).

Having a vegetable garden contributes to being a good neighbor. Many vegetables, such as zucchini and other varieties of summer squash, are prodigious producers, providing the gardener with a cornucopia of abundance that is far more than any family can consume. Friends and neighbors invariably appreciate receiving some of this abundance. Sharing with others vegetables raised in one's garden is a good way to build community.

It is said that time spent fishing is not time that is subtracted from one's life. The same is true of gardening.

URBAN GARDENING

Some assume that one has to live on a farm or out in the country in order to have a vegetable garden. However, that is not the case. In fact, urban gardening has been growing in popularity in recent years.

In a book entitled *City Bountiful: A Century of Community Gardening in America*, Laura J. Lawson, Dean of Agriculture and Urban Programs at Rutgers University, observes:

> Children's delighted faces as they pull up the carrots they had planted from seeds, neighbors working together to transform a vacant lot into a garden, mouth-watering vegetables just picked from the earth—these are what come to the minds of many when they think of community gardens. Such images will likely cause experienced gardeners to nod knowingly, having enjoyed both the work and the rewards of the garden.[40]

Lawson speaks from personal experience, having served as coordinator of the Berkeley Youth Alternatives (BYA) Community Garden Patch while completing degrees in landscape architecture and environmental planning at the University of California, Berkeley.

Lawson notes that urban garden programs include a wide variety of cooperative enterprises that provide resources and space for those living in urban areas to grow vegetables and flowers. Some are community gardens in which participants have their own plots but share in the garden's management. Other types of urban gardens include children's gardens, job-training gardens, horticultural therapy gardens and many other types of gardens.[41] Lawson suggests that there are three "significant recurring themes" in these various types of gardens:

> [One of these themes is that] urban garden programs provide a participatory experience that connects people living in cities, especially children, to the soil and plant and animal life. . . . This reconnection with nature is often associated with improving social and psychological health. . . . A second theme is the association of urban gardens with education. Through gardening, one learns not only practical skills associated with gardening—the steps necessary to nurture seed to fruit—but also the civic-mindedness to nurture a community open space. . . . A third theme is the portrayal of gardens as a democratic space and gardening as an activity that brings diverse groups together in mutual self-interest.[42]

There is a related point that merits underscoring. As Leopold serves to remind us, environmental awareness and responsibility are best learned experientially—from the bottom on up, rather than from the top on down. Part of the insightfulness of urban gardening programs is that they help children and others connect with nature, expanding their environmental awareness. They also learn responsibility by seeing first-hand the dividends in the form of good things to eat that are the result of the work that they do.

As Lawson notes, there are many different types of urban gardens. Here are a few examples:

- The Bradner Gardens Park, which is located in the Mount Baker neighborhood in Seattle, an ethnically diverse neighborhood, has sixty-one 200–square foot plots for gardening. The community garden was established in 1987 to provide gardening space for Mien immigrants from Laos. Other residents of the Mount Baker neighborhood joined the endeavor. The site is owned by the Seattle Department of Parks and Recreation, which maintains the property. A volunteer committee comprised of representatives of affiliated organizations manages the gardens and provides gardening advice to those with garden plots. Bradner Gardens Park is one of more than sixty community gardens scattered throughout the city of Seattle. Any Seattle resident may apply for a garden plot.[43]
- The Chicago Harvest Garden Program, a three-season organic gardening program for children ages 6–12, teaches children how to plant and maintain an edible garden in a park in their neighborhood. The program, which is managed by the Chicago Park District's Culture, Arts & Nature Department, is available in sixteen parks spread throughout the city. The sixteen-week curriculum includes the basics of gardening, as well as lessons in nutrition, cooking, and environmental stewardship. During the last week of the program, the Harvest Festival brings together program participants from all of the parks to show their produce and talk about what they accomplished.[44]
- The BYA Community Garden Patch (with which Lawson was involved) had its beginning in 1993 when BYA Executive Director Niculia "Nikki" Williams, dismayed by seeing children arrive at the youth center with fast food breakfasts, decided to start a small garden. That was the seed for an idea that three years later had grown into the BYA Community Garden Patch. The heart of Garden Patch program is the youth market garden, which employs up to six young people, depending on funding. They work fifteen hours a week under the direction of a garden supervisor. The youth sell the produce they grow at farmer's markets and to local restaurants. The BYA Garden Patch is also involved in other youth-oriented activities. For example, one year students from a school for developmentally disabled youth maintained several plots.[45] The BYA Community Garden Patch is a program of BYA, a community-based organization which for over forty years has been serving the children and families of Berkeley and the surrounding cities in Alameda County and Contra Costa County. BYA's mission emphasizes three core areas: education, health and well-being, and economic self-sufficiency.[46]
- Nearly 300 community gardens are scattered throughout the city of Minneapolis. About a fifth of these are on vacant city-owned lots. In addition, the city has more than 100 vacant lots available for qualifying groups to lease. Qualifying community garden groups must have a non-profit or neighborhood sponsor. The sponsors are required to have liability insu-

> rance that covers the lot. All of the leasable city-owned lots have had soil tests for lead and other potential contaminants. Minneapolis Solid Waste and Recycling offers low-cost or no-cost compost to registered community gardens, with the provision that community gardens that have never received compost, have poor soil quality or are start-up gardens receive priority. The Water Works Permit Office issues garden water permits to community gardens and urban farms to access a specific fire hydrant for their garden or farm use if there is no other water option available. The group applying for the water permit must have legal documentation from the owner of the lot giving permission to use the lot for a garden. The permits are issued seasonally. Water use is metered during the growing season with the sponsoring group billed for the water.[47]

Community gardens, it should be added, are not the only option for gardeners living in urban areas. Parsley, chives, basil, and other herbs can be grown in containers on a patio or a balcony that receives several hours of sunlight each day. So also can tomatoes. Roma tomatoes are a good choice for container cultivation, though with proper care, Early Girl and other varieties can also be grown. Whether being grown in a conventional garden or in a large container on a balcony or patio, the vines of tomatoes need support. This can be accomplished either by purchasing a metal tomato cage or by placing a three-foot wooden stake in the center of the soil in the container and then using twine to tie the vines to the stake.

CITIZENSHIP AND STEWARDSHIP

While Leopold is correct in insisting that humankind ought to eschew the role of conqueror of the land community, it does not follow that humans are simply members of the land community who are interchangeable with all other members of the land community. Rather, humans possess levels of intelligence and technological abilities which exceed those of other members of the land community. This is not to suggest, as so many have, that this makes humans superior to other species—an assumption which historically has often been very harmful to other species and otherwise destructive of ecosystems. Rather, it is to suggest that humans have special responsibilities with respect to other members of the land community. This is where the notion of stewardship comes into play.

Saying that humans, in general, and private property owners, in particular, have special responsibilities of stewardship need not—and ought not—morph into an anthropocentric view of humankind's relationship to the non-human members of the land community. This is not an implicit license to exploit. Rather, this special responsibility should result in "a conviction of individual responsibility for the health of the land."[48] This includes following appropriate conservation measures and otherwise protecting the well-being of the land. In short, we are citizens of the land community with stewardship responsibilities, as well as plain members of the land community.

CONSERVATION EASEMENTS

Much has happened, of course, since Leopold wrote *A Sand County Almanac* in the late 1940s—some of it positive, some of it not so positive. One of the positive developments is the advent of conservation easements, which enable conservation-minded property owners with love for the land to protect their property in perpetuity from sub-divisions, commercial development, surface mining, and other activities which would have a negative environmental impact on the land. As of April 22, 2019, there were 158,168 conservation easements in the United States, covering more than 27 million acres.[49]

While there are many different variations of conservation easements, they all essentially work as follows: conservation-minded landowners who choose to do so (conservation easements are always voluntary) give up specified rights, such as the right to subdivide their property and turn it into a housing development, while retaining all other rights related to the use of their land. The rights that landowners choose to give up in order to protect their land are donated to a private not-for-profit conservation organization, which then functions as a conservation guardian of the property on which the easement has been placed. This easement is attached to the property in perpetuity, which means that if the landowner sells the land to someone else, or if the land is inherited by the landowner's children or other individuals, the restrictions intended to protect and preserve the land continue to be in effect and cannot be canceled by the new owners.[50]

The Montana Land Reliance, a private not-for-profit organization which since its founding in 1978 has conserved more than 1 million acres in nearly 900 easements spread throughout the state of Montana, is fairly typical of the conservation groups that draft and monitor conservation easements. The organization's website states, "A conservation easement is the legal glue that binds a property owner's good intentions to the land in perpetuity. Donors of conservation easements retain title to their property. They grant conservation easements to protect their land from inappropriate development. A conservation easement runs with the title to the property regardless of changes in future ownership."[51]

Uses of the property that are restricted by Montana Land Reliance conservation easements, as agreed to by the landowner, are often the following:

- subdivision for residential development,

- construction of non-agricultural buildings,
- non-agricultural commercial activities,
- dumping non-compostable or toxic waste, and
- surface mining

Rights retained by the landowner include, but are not limited to, the following:

- continued agricultural and silvicultural (forestry) use,
- constructing buildings, fences, and other measures for agricultural uses,
- landowner control of access to the property,
- wildlife and fisheries protection, restoration and enhancement projects, and
- any and all other uses not specifically prohibited.[52]

And so building a Wal-Mart store is out of the picture. On the other hand, living on the land and using the land for agricultural purposes, while taking care of the land in the way that Leopold believed should be done, is not out of the picture. Indeed, that is precisely what conservation easements are intended to help ensure.

There are some potential tax benefits related to conservation easements. Conservation easements, by prohibiting subdivision and housing projects, often reduce the value of the property on which the easement is placed. The difference in the value of the property before the easement and after the easement, as determined by a qualified and licensed appraiser, counts as a charitable contribution to the conservation organization drafting and monitoring the agreement and, within certain limits, may be listed as a charitable contribution on the landowner's federal income tax form.[53] The amount that the landowner saves in taxes, of course, depends on her or his tax bracket. It is, however, always less than the amount by which the value of the land has diminished since even the highest tax rates are only a percentage of adjusted gross income.

The most important reason to place a conservation easement on property one owns, though, is not to reduce what one pays in taxes. Rather, it is to practice good stewardship by protecting the land one owns from development and other activities that would harm it.

STORIES OF HOPE

There are, of course, a far greater number of environmental issues than can be considered in any single volume. In order to keep the discussion as focused as possible, particular attention in the following chapters is devoted to three sets of environmental issues: (1) protection of endangered species, (2) the sustainable agriculture movement, and (3) recycling. In all three cases, the issues will be set in a historical context, with narrative used to help frame the discussion. All of these stories are set against the backdrop of huge environmental catastrophes. All of them are also stories of hope, which is why they were chosen to be included in this volume.

A PERSONAL POSTSCRIPT

I grew up on a small family farm in a mountain valley in northwestern Montana. We had cattle, both beef cattle and milk cows, chickens and, part of the time, pigs. Only about eighty acres on the 240–acre family farm were under cultivation. The rest was timberland used for pasture. Overall, the farm was not very productive. We struggled to make ends meet.

While growing up, I wore hand-me-down clothes. We did not have television, not because television had not been invented but because our family lacked the money to buy a television set. In retrospect, growing up without television was a blessing—a blessing that few today experience.

Though we had little money, we still had a very good life. We always had plenty of food to eat, most of it produced on the farm. It was healthy food—food that was not laced with all sorts of preservatives and other chemicals that do not belong in food. My mother baked whole wheat bread using flour my father ground from hard red winter wheat he raised on the farm. It was good flour, which made for good bread. Indeed, one cannot make good bread without having good flour from which to make it. I did not realize at the time how privileged I was to have such nutritious bread and the other wholesome, healthy food that came from the farm.

Many today equate quality of life with levels of affluence. That is misleading. There are many factors which are far more important in determining quality of life, including having wholesome, healthy food to eat. Coming to the realization that money is not everything is one of the most important milestones of awareness one can experience—a milestone which is difficult for many to realize today in an era in which mass advertising promotes consumption, much of it wasteful, while various cultural factors encourage mindless pursuit of wealth.

While I was in college (I had the good fortune of being awarded a full-ride scholarship for college), the valley where our farm is located started changing in ways that were not for the better. Affluent people from other parts of the country discovered the valley and wished to have homes there, some of them vacation homes, some of them retirement homes. Developers who cared only about making as much money as possible were happy to accommodate the affluent and became wealthy themselves in the process.

Farmers struggling to make ends meet were tempted by the prices developers offered them for their land. Some sold their land to developers and lived off the money they received as long as it lasted. What they were paid for their land was far less than what developers made by subdividing it and building expensive homes on it. In the span of just a few years, parts of the valley looked like, as one of the locals put it, "a suburb of Los Angeles."

Some, including my father, refused to sell. Using language that was not particularly gracious, he bluntly told realtors to get off his land when they stopped by to see if he might be interested in selling. It was because of my father's obstinacy that I came to own part of the family farm. When my parents passed away, my older sister, my younger brother, and I each inherited eighty acres of the family farm—beautiful land that we could not have afforded to buy on the open market.

My brother got the home place with the farm buildings, which was fine by me. They are old buildings which are expensive to maintain. (My childhood room is still there in the old farmhouse, pretty much the way it was when I left to go to college.) The eighty acres I inherited are mostly wooded with about fifteen open acres which had been used to raise hay but are now pasture for horses.

One of the first things I did when I inherited my share of the family farm was place a conservation easement on it, which protects the land in perpetuity from being subdivided for development. There are other restrictions as well that would go with the property should we ever sell it (which we have no intention of doing), including provisions to protect the wooded areas that are habitat for wildlife—provisions that have had the practical impact of making our land a refuge for wildlife, which is exactly what we want it to be. We do not allow hunting on our property. A particularly significant provision of the conservation easement protects a wooded corridor that serves as part of a passageway for grizzly bears and other wild animals traveling from the Mission Mountains to the Swan Range of the Rocky Mountains. Wildlife biologists tell us that wildlife mobility is important if genetic diversity is to be maintained.

Because our land is not particularly good farmland, the primary value of the land when I inherited it was its potential for development. Once my brother and I put conservation easements on our property, that potential was no longer there, which substantially reduced the market value of our property. That was acceptable to us. As noted, there are many things that are far more important than money. Protecting the land we are privileged to own is one of them.

My wife and I built a two-bedroom log cabin on my share of the family farm, a project that stretched out over seven years. Because I enjoy building things, I did much of the work on the cabin myself, with my wife helping with staining and other aspects of the project when her schedule allowed. A high-efficiency wood-burning stove and a wood-burning fireplace heat the cabin. There is an abundance of firewood to saw and split on our land, most of it coming from trees that have blown over.

Our land is rich with wildlife—white-tailed deer, wild turkeys, black bears, an occasional grizzly bear, coyotes, badgers, an occasional red fox, red-tailed hawks, bald eagles, osprey, ravens, great horned owls, nighthawks, ruffed grouse, Canada geese, mallards, belted kingfishers, downy woodpeckers, pileated woodpeckers, northern flickers, western tanagers, mountain chickadees, juncos, black-chinned hummingbirds, and many other species. There is also an abundance of wild flowers, among them wild roses, blue lupine, lavender asters, sego lilies, ox-eye daisies, yarrow, blue bells, and heartleaf arnica.

We are taking measures to ensure the health of the wooded areas on our land. Arne Brosten, a cousin who for years worked for the U.S. Forest Service, put together a forest management plan for us, which involves some selective logging. At our request, the conservation easement prohibits clear-cutting, a provision intended to protect the wooded areas from being destroyed. However, the easement does allow selective logging to improve the health of the forest by removing diseased trees (quite a few of the spruce trees on our land have root rot, which leaves them vulnerable to blowing over in strong winds) and preventing overcrowding (in some cases there are two trees growing from the same stump, which is not good for either of them).

Because highly-mechanized forms of logging often tear up the land, we had a strong preference for low-impact logging. Jerry Benson and his son, Justin, turned out to be exactly what we were looking for. The last horse loggers in the valley, they have a beautiful team of Belgian draft horses, which they use to skid the logs to the landing prior to being loaded onto the truck that hauls them to the saw mill. Speck and Pet each weigh more than a ton. Notwithstanding their size, they are poetry in motion as they work together skidding logs, their movement completely synchronized. Speck and Pet take great pride in their work.

When one cares about the land one is privileged to own, there is a special bond that forms. As did Leopold with respect to the run-down farm he purchased, I have a deep-seated love for our land and for the wildlife, wildflowers, and trees that inhabit it. The land is part of me, and I am part of it.

NOTES

1. Henry David Thoreau, "The Pond in Winter" in *Walden and Other Writings of Henry David Thoreau*, ed. Brooks Atkinson (New York: The Modern Library, 1950), 254.
2. Thoreau, "Spring" in *Walden*, 283.
3. Aldo Leopold, *A Sand County Almanac with Essays on Conservation from Round River* (New York: Ballantine Books, 1966), xviii.
4. Ibid., 37.

5. Ibid., 54–55.

6. Ibid., 57.

7. Quoted by Barry Yeoman in "Why the Passenger Pigeon Went Extinct," *Audubon Magazine*, May–June 2014, at https://www.audubon.org/magazine/may-june-2014/why-passenger-pigeon-went-extinct (last accessed April 30, 2019).

8. United States Fish and Wildlife Service, "Time Line of the American Bison" at http://www.fws.gov/bisonrange/timeline.htm (last accessed April 22, 2019).

9. Eckardt C. Beck, "The Love Canal Tragedy," *EPA Journal*, January 1979, at https://archive.epa.gov/epa/aboutepa/love-canal-tragedy.html (last accessed April 22, 2019).

10. Donald G. McNeil, Jr., "Upstate Waste Site May Endanger Lives," *New York Times*, August 2, 1978.

11. Montana Bureau of Mines and Geology, "Environmental Assessment—Berkeley Pit" at https://www.mbmg.mtech.edu/env/env-berkeley.asp (last accessed December 9, 2016). The Anaconda Copper Mining Company opened the Berkeley Pit in 1955. The Atlantic Richfield Company, which purchased the Anaconda Copper Mining Company in 1977, continued mining operations at the Berkeley Pit until 1982.

12. Ben Guarino, "Thousands of Montana Snow Geese Die after Landing in Toxic, Acidic Mine Pit," *Washington Post*, December 7, 2016. See also Brian James Leech, *The City That Ate Itself: Butte, Montana and Its Expanding Berkeley Pit* (Reno: University of Nevada Press, 2018).

13. "Nutrient Pollution—Eutrophication," NOAA Ocean Service Education at http://oceanservice.noaa.gov/education/kits/estuaries/media/supp_estuar09b_eutro.html (last accessed April 22, 2019).

14. "Hypoxia in the Northern Gulf of Mexico," Louisiana Universities Marine Consortium at http://www.gulfhypoxia.net/overview/ (last accessed December 16, 2016).

15. The origins of this quotation, which appears in different versions, are somewhat obscure, though it is believed to be Native American (The Quote Investigator at http://quoteinvestigator.com/2011/10/20/last-tree-cut/ [last accessed April 22, 2019]).

16. Quoted by Hans Huth in *Nature and the American: Three Centuries of Changing Attitudes* (Berkeley: University of California Press, 1957), 5.

17. John Fraser Hart, "Change in the Corn Belt," 53–55, at http://www.geo.hunter.cuny.edu/courses/geog347/articles/change_belt.pdf (last accessed April 22, 2019).

18. Aldo Leopold, *A Sand County Almanac with Essays on Conservation from Round River* (New York: Ballantine Books, 1966), 262–263.

19. "About the Leopold Center," The Leopold Center for Sustainable Agriculture at Iowa State University at https://www.leopold.iastate.edu/about/leopold-center (last accessed April 22, 2019).

20. See, for example, Joseph D. Cornell and Michelle Miller, "Slash and Burn," *Encyclopedia of Earth*, ed. Cutler J. Cleveland (Washington, D.C.: National Council for Science and the Environment, 2011), at http://www.eoearth.org/article/Slash_and_burn (last accessed May 7, 2018).

21. Paul VI, *Populorum Progressio*, 18, at http://w2.vatican.va/content/paul-vi/en/encyclicals/documents/hf_p-vi_enc_26031967_populorum.html (last accessed April 30, 2019).

22. Leopold, *A Sand County Almanac*, 240.

23. Ibid., 239.

24. J. Baird Callicott, *In Defense of the Land Ethic: Essays in Environmental Philosophy* (State University of New York Press, 1989), 136–138. Callicott refers to the Genesis 2 story of creation as "J-theism," the label referring to a widely accepted theory about the origin of the text.

25. Bruce M. Metzger and Roland E. Murphy, eds., *The New Oxford Annotated Bible*, (Oxford University Press, 1991), 3.

26. Callicott, *In Defense of the Land Ethic*, 155.

27. Ibid., 138.

28. Ibid., 150–151.

29. Ibid., 154.

30. Francis, *Laudato Si': On Care for Our Common Home*, May 24, 2015, paragraph 11, at http://w2.vatican.va/content/francesco/en/encyclicals/documents/papa-francesco_20150524_enciclica-laudato-si.html (last accessed April 30, 2019).

31. Leopold, *A Sand County Almanac*, 6–7.

32. Ibid., 54–55.

33. Ibid., 57.

34. Stephen R. Kellert et al., *The Nature of Americans: Disconnection and Recommendations for Reconnection*, Nature of Americans.org, April 2017, 3–6, 87–91, 267–272 at https://natureofamericans.org/sites/default/files/reports/Nature-of-Americans_National_Report_1.3_4–26–17.pdf (last accessed April 18, 2019).

35. Lincoln R. Larson *et al.*, "Outdoor Time, Screen Time, and Connection with Nature: Troubling Trends Among Rural Youths?" *Environment and Behavior*, 20 October 2018, 1–26 at https://journals.sagepub.com/doi/pdf/10.1177/0013916518806686 (last accessed April 18, 2019).

36. Kellert et al., *The Nature of Americans*, 291–300.

37. National Park Service, "Visitation Numbers" at https://www.nps.gov/aboutus/visitation-numbers.htm (last accessed April 18, 2019).

38. Henry David Thoreau, "The Bean Field" in *Walden and Other Writings of Henry David Thoreau,* ed. Brooks Atkinson (New York: The Modern Library, 1950), 140. In Greek mythology, Antæus, who was the son of Poseidon, the sea god, and Gaea, the earth goddess, had his strength renewed whenever he touch the earth.

39. Nathaniel Hawthorne, *Mosses from an Old Manse* (Columbus: Ohio State University Press, 1974), 13–14.

40. Laura J. Lawson, *City Bountiful: A Century of Community Gardening in America* (Berkeley: University of California Press, 2005), 1.

41. Ibid., 3.

42. Ibid., 5–7.

43. "Bradenton Gardens Park" at http://www.bradnergardenspark.org/index.html (last accessed April 22, 2019). See also Jeffrey Hou, Julie M. Johnson, and Laura J. Lawson, *Greening Cities, Growing Communities: Learning from Seattle's Urban Community Gardens* (University of Washington Press, 2009), 107–122.

44. Chicago Park District, "Community Gardens" at https://www.chicagoparkdistrict.com/parks-facilities/community-garden (last accessed April 22, 2019).

45. Lawson, *City Bountiful*, 274–281.

46. Berkeley Youth Alternatives, "Our Mission" at http://www.byaonline.org/about-us/our-mission/our-mission (last accessed April 22, 2019).

47. City of Minneapolis, "Homegrown Minneapolis Community Gardens" at http://www.minneapolismn.gov/sustainability/homegrown/WCMS1P-129871 (last accessed April 22, 2019).

48. Leopold, *Sand County Almanac,* 258.

49. *National Conservation Easement Data Base* at https://www.conservationeasement.us/ (last accessed 22 April 2019).

50. The Nature Conservancy, "Conservation Easements" at http://www.nature.org/about-us/private-lands-conservation/conservation-easements/all-about-conservation-easements.xml (last accessed April 22, 2019).

51. Montana Land Reliance, "Conservation Easements" at http://mtlandreliance.org/conservation-easements/ (last accessed April 22, 2019).

52. Ibid.

53. Ibid.

Part I

Extinction and Endangered Species

Chapter One

Ode to a Pigeon

In autumn of 1813, John James Audubon, a gifted artist and naturalist whose book, *The Birds of America*, is one of the most celebrated works ever published, was traveling by horseback from his home in Henderson, Kentucky, to Louisville, Kentucky, a town by the Ohio River founded during the American War for Independence.[1] Audubon stopped his horse when he saw a huge flock of birds coming straight at him. What he witnessed was an incredible spectacle that defies comprehension.

The huge flock of passenger pigeons swept overhead from one edge of the sky to the other. Audubon later wrote, "The air was literally filled with pigeons. The light of noon-day was obscured as by an eclipse, the dung fell in spots, not unlike melting flakes of snow; and the continued buzz of the wings had a tendency to lull my senses to repose."[2] When Audubon arrived in Louisville at the end of the day, the pigeons were still flying overhead. The banks of the Ohio River were crowded with men and boys shooting at the pigeons, with piles of dead pigeons at their feet. The flight continued throughout the night and into the next day and the day after that.[3] For more than a week, residents of Louisville ate no meat other than the flesh of the pigeons they had shot.[4] Audubon later estimated that the flock was comprised of 1,115,136,000 birds.[5]

The huge flock of passenger pigeons that Audubon observed was not unique. An 1855 report from Columbus, Ohio, described a "growing cloud" that blotted out the sun as it approached the city. "Children screamed and ran for home. Women gathered their long skirts and hurried for the shelter of stores. Horses bolted. A few people mumbled frightened words about the approach of the millennium, and several dropped on their knees and prayed." Two hours later, after the flock had passed over, "the town looked ghostly in the now-bright sunlight . . ."[6]

In 1871, an estimated 136 million breeding birds occupied 850 square miles of forestland in Wisconsin.[7] Hunters, coming upon the morning exodus of adult males, were so overwhelmed by the sound that some of them dropped their guns. A newspaper in Fond du Lac said of that encounter, "Imagine a thousand threshing machines running under full headway, accompanied by as many steamboats groaning off steam, with an equal quota of [railroad] trains passing through covered bridges—imagine these massed into a single flock, and you possibly have a faint conception of the terrific roar."[8] Passenger pigeons were believed to comprise 25 to 40 percent of the total U.S. land bird population.[9]

Passenger pigeons (*Ectopistes migratorius*) were handsome birds. Half again the size of mourning doves,[10] they had slate blue heads. The backs of their necks were an iridescent bronze that that changed to green or purple in different light conditions. Their backs were slate gray tinged with olive brown while their throats and breasts were pale cinnamon. They had two central tail feathers that were brownish gray with the rest of their tail feathers white. Their beaks were black, their legs and feet red. The colors of the females were not as vivid as those of the males.[11]

Passenger pigeons were as graceful in flight as they were handsome. Audubon observed, "When an individual is seen gliding through the woods and close to the observer, it passes like a thought, and on trying to see it again, the eye searches in vain; the bird is gone."[12]

Less than a half century after 136 million breeding birds occupied 850 square miles of forestland in Wisconsin, there were no longer any passenger pigeons. In the early afternoon of September 1, 1914, as a war of horrific proportions was raging half a world away, Martha, the last surviving passenger pigeon, died in her cage at the Cincinnati Zoological Gardens. She had never lived in the wild. She is believed to have been hatched at Chicago's Brookfield Zoo. Her mother may have lived earlier at the Milwaukee Zoo. By 1900, these three zoo populations were all that was left of a species that had just a few years earlier numbered in the billions. Fourteen years later, these three small remnants were gone.[13]

As the end of her life approached, Martha lived alone in her cage. Her wings drooped and she trembled. Zoo keepers had to rope off her cage to keep visitors from throwing sand at her to try to make her move. After she died, her body was packed in ice and shipped to the Smithsonian Institution in Washington, D.C., where taxidermists skinned her and stuffed her so that she could be put on display. She can be seen there today.[14]

Some say deforestation or an imported avian disease might have contributed to the demise of passenger pigeons.[15] Most agree, however, that the major factor was the wholesale slaughter of these birds by another species, namely *Homo sapiens*. Some people viewed shooting passenger pigeons as a sport. Others saw economic opportunity in killing them in massive numbers and shipping them by fast train to eastern cities, where their flesh was considered to be a tasty addition to the menu. In 1871, the *Kilbourn City Mirror*, published in what today is Wisconsin Dells, reported, "Hardly a train arrives that does not bring hunters or trappers . . . Hotels are full, coopers are busy making barrels, and men, women and children are active in packing the birds or filling the barrels. They are shipped to all places on the railroad, and to Milwaukee, Chicago, St. Louis, Cincinnati, Philadelphia, New York, and Boston."[16]

Commercial hunters shot the pigeons, used nets to trap them, or torched their roosts.[17] Some hunters would lure the birds to feed on alcohol-soaked grain; when the birds were inebriated, they were easy to catch. Other hunters would suffocate large numbers of nesting or roosting birds with grass or sulfur fires lit beneath the trees.[18] As noted in the Prologue to this volume, Potawatomi tribal leader Simon Pokagon, seeing what was happening, wondered what type of divine punishment would be "awaiting our white neighbors who have so wantonly butchered and driven from our forests these wild pigeons, the most beautiful flowers of the animal creation of North America."[19]

Michigan was one of the last strongholds of passenger pigeons. On just one day—July 23, 1860—235,200 birds were shipped east from Grand Rapids. In 1869, Van Buren County sent 7,500,000 birds east. Oceana County did not do quite as well. In 1874 the county sent 1,000,000 birds east, which increased to 1,600,000 birds two years later.[20] In 1878, a single hunter shipped 3 million Michigan birds to eastern cities. By 1889, passenger pigeons were extinct in the state.[21]

The story was essentially the same in other areas that had once been home to hundreds of millions of passenger pigeons. In 1910, the American Ornithologists' Union began a two-year search for passenger pigeons. They offered rewards totaling more than $2,000 to be paid to anyone who could find a passenger pigeon living in the wild. The reward money went unclaimed.[22]

The Wisconsin Society for Ornithology erected a monument to the passenger pigeon in Wyalusing State Park in Wisconsin on a bluff overlooking the Mississippi River, a monument which was dedicated May 11, 1947. A plaque on the monument states, "This species became extinct through the avarice and thoughtlessness of man."[23] In an essay entitled "On a Monument to the Pigeon" written for the dedication, Aldo Leopold observes:

> We grieve because no living man will see again the onrushing phalanx of victorious birds, sweeping a path for spring across the March skies, chasing the defeated winter from all the woods and prairies of Wisconsin.
>
> Men still live who, in their youth, remember pigeons. Trees still live who, in their youth, were shaken by a living wind. But a decade hence only the oldest oaks will remember, and at long last only the hills will know.[24]

NOTES

1. The town was named after Louis XVI of France, who, as a result of patient and skillful negotiations by Benjamin Franklin, had agreed to help the American side of the conflict.
2. Quoted by William Souder in "100 Years After Her Death, Martha, the Last Passenger Pigeon Still Resonates," *Smithsonian Magazine*, October 2014 at http://www.smithsonianmag.com/smithsonian-institution/100-years-after-death-martha-last-passenger-pigeon-still-resonates-180952445/?no-ist (last accessed February 2, 2017)
3. Ibid.
4. Edward Howe Forbush, "Extinction of the American Passenger Pigeon" at http://www.wildbirds.org/apidesay.htm (last accessed September 22, 2015). The essay originally appeared in a publication of the Massachusetts Board of Agriculture.
5. "Passenger Pigeon: Demography and Populations," *Birds of North America Online* at http://bna.birds.cornell.edu/bna/species/611/articles/demography?searchterm=Passenger%20Pigeon (last accessed June 15, 2019).
6. Quoted by Barry Yeoman in "Why the Passenger Pigeon Went Extinct," *Audubon Magazine*, May–June 2014, at https://www.audubon.org/magazine/may-june-2014/why-passenger-pigeon-went-extinct (last accessed June 15, 2019).
7. Souder, "100 Years After Her Death, Martha, the Last Passenger Pigeon Still Resonates."
8. Quoted by Yeoman in "Why the Passenger Pigeon Went Extinct."
9. Walter Beacham, ed., *The Official World Wildlife Fund Guide to Extinct Species of Modern Times*, Volume I (Beacham Publishing Corp., 1997), 182
10. Souder, "100 Years After Her "Death, Martha, the Last Passenger Pigeon Still Resonates."
11. Beacham, ed., *The Official World Wildlife Fund Guide to Extinct Species of Modern Times*, 182.
12. Ibid., 185.
13. Souder, "100 Years After Her "Death, Martha, the Last Passenger Pigeon Still Resonates."
14. Ibid.
15. Beacham, ed., *The Official World Wildlife Fund Guide to Extinct Species of Modern Times*, 184.
16. Quoted by Yeoman in "Why the Passenger Pigeon Went Extinct."
17. Ibid.
18. Beacham, ed., *The Official World Wildlife Fund Guide to Extinct Species of Modern Times*, 184.
19. Quoted by Yeoman in "Why the Passenger Pigeon Went Extinct."
20. Clive Ponting, "Species Extinction: The Story of the Passenger Pigeon" at http://www.eco-action.org/dt/pigeon.html (last accessed June 15, 2019). The selection is excerpted from Ponting's book, *A Green History of the World* (Penguin Books, 1992), 168–170.
21. Beacham, ed., *The Official World Wildlife Fund Guide to Extinct Species of Modern Times*, 184.
22. "Passenger Pigeon: Demography and Populations," *Birds of North America Online*.
23. "On a Monument to the Pigeon," by Aldo Leopold, *Birdwatching Daily* at https://www.birdwatchingdaily.com/blog/2014/09/01/monument-pigeon-aldo-leopold/ (last accessed June 10, 2019).
24. Aldo Leopold, "On a Monument to the Pigeon" in *A Sand County Almanac* (New York: Ballantine Books, 1966), 116.

Chapter Two

From Thundering Herds to But a Few

They are huge animals. Males, which are larger than females, can be more than six feet high at the front shoulder. From head to tail, they can measure more than twelve feet. Mature males often weigh as much as a ton. Their heads and front shoulders are covered with thick brown fur, with the hair on their backs and hind quarters somewhat shorter. Their winter coats are so thick that snow can cover their backs without melting. Both males and females have curved horns that point upward, with the horns of the male somewhat larger. Their eyesight is poor, but they have acute hearing and sense of smell.[1]

Though just about everyone refers to them as "buffalo," they are only remotely related to the water buffalo of Southeast Asia and other true buffalo. It is more accurate to refer to them as "bison," or, somewhat more formally, by their scientific name, *Bison bison*. There are two recognized subspecies of the North American bison—*Bison bison bison* (the plains bison) and *Bison bison athabascae* (the wood bison).[2]

Bison were the source of sustenance and just about everything else needed by the Native American tribes living on the Great Plains in what is now the United States and on the prairies of what is now Canada—tribes that included the Blackfoot, Cheyenne, Comanche, Crow, Lakota, and many others. Bison meat, which was dried and saved for winter use, often combined with fat and berries to form pemmican, comprised a major portion of their diets. Bison hides stretched over poles tied together near the top formed the tipis which protected them from inclement weather. The hides also were the source of leather for the clothing they wore[3] and the moccasins that protected their feet. Bison robes kept them warm in the winter and were the bedding in their tipis on which they slept. Bison sinews provided the strings for their bows and the thread they used when making their clothing. Bison bones were formed into tools and other useful items. In short, just about everything from the bison that could be used was used. And just about everything that sustained the way of life of the plains tribes, other than the poles for their tipis, the wood for their bows, and the flint and obsidian used to make the points for their arrows and spears, came from bison. As one author put it, "Plains people manufactured many artifacts from the hides, horns, and hoofs of the animal, subsisted on its flesh, wove legends about the herds, and accepted the bison as a spiritual force of their Great Plains world."[4]

Prior to the acquisition of horses, Native Americans often hunted on foot, trying to get close enough to bison and other game animals to bring them down with a well-placed arrow or the thrust of a spear. It was an endeavor fraught with danger since a bison that was not brought down could—and often did—charge the hunter.[5]

Though also fraught with danger, plains tribes developed other techniques for harvesting the bison they needed to sustain their way of life. Bison jumps were one of the most effective of these techniques. The location had to be carefully chosen. It needed to be near a grassy area where bison would be grazing. It needed to be near a cliff over which stampeding bison could be driven. And it needed to be near a river or some other source of water for processing the carcasses once the bison had been driven over the cliff.

A number of these sites still exist in Canada and in the United States. Today, several are parks which those interested in the history of the plains tribes may visit. These include Dry Island Buffalo Jump Provincial Park in central Alberta, Canada,[6] Madison Buffalo Jump State Park near Three Forks, Montana,[7] and First Peoples Buffalo Jump State Park near Great Falls, Montana,[8] which the United States National Park Service officially designated as a National Historic Landmark on August 4, 2015.[9]

Though there were minor variations from site to site and from tribe to tribe, the basic game plan was essentially the same. Cairns of piled stones, with brush often spread out between them, would form drive lines designed to channel the bison toward the jump—drive lines that in some cases were more than a mile long. The *ahwa waki* (buffalo runner) played a key role. Dressed in a robe made of the skin of a calf and often wearing a cap with bison ears and horns, the *ahwa waki*, a young, very athletic male member of the tribe, would position himself ahead of the herd in the direction of the jump. When the lead cow (a herd of bison typically follow a lead cow, usually one of the older and wiser cows in the herd) started moving toward the *ahwa waki*, he would move in the direction of the jump. Other hunters wearing wolf hides would spring up from behind the cairns to spur the bison on.[10] In the May 29, 1805, entry in his journal, Captain Meriwether Lewis, who came across a jump site by the Missouri River while on his journey of discovery with Second Lieutenant William Clark and a select group of volunteers, tells the rest of the story:

> [T]he disguised indian or decoy has taken care to place himself sufficiently nigh the buffaloe to be noticed by them when they take to flight and runing before them they follow him in full speede to the precipice, the cattle behind driving those in front over and seeing them go do not look or hesitate about following untill the whole are precipitated down the precepice forming one common mass of dead an[d] mangled carcases: the decoy in the mean time has taken care to secure himself in some cranney or crivice of the clift which he had previously prepared for that purpose.[11]

Lewis, in somewhat of an understatement, added that "the part of the decoy I am informed is extreamly dangerous, if they are not very fleet runers the buffaloe tread them under foot and crush them to death, and sometimes drive them over the precipice also, where they perish in common with the buffaloe."[12]

When the plains tribes acquired horses, the methods of hunting bison changed. Hunters on horses would ride up alongside bison and either shoot them with arrows or guns or thrust a spear into them.

When Europeans first came to North America, the estimated bison population was 30 to 60 million.[13] Colonists originally settled along the eastern coast of what is now the United States, but as their descendants and an ever growing stream of immigrants from the Old World moved inland, bison were pushed back to the Great Plains. By 1730, all the wild bison were gone from Virginia. Frontiersman Daniel Boone wrote in his diary that he hunted bison in North Carolina until they "became scarce" in the late 1760s. Shortly thereafter, bison could no longer be found in the Carolinas or eastern Georgia.[14] By 1802, they were gone from Ohio.[15]

But though bison no longer roamed in the eastern part of the North American continent, on the Great Plains they continued to be found in abundance, as evidenced by numerous references to large herds of buffalo in the *Journals of the Lewis and Clark Expedition*. On July 11, 1806, Lewis wrote that when he "arrived in sight of the white-bear Islands the missouri bottoms on both sides of the river were crouded with buffaloe." He added, "I sincerely beleif that there were not less than 10 thousand buffaloe within a circle of 2 miles around that place."[16] Similar observations were made on subsequent days. On July 18, he reported that they "passed immence herds of buffaloe . . . for about 12 miles it appeared as one herd . . . ."[17] And on July 27, he wrote that "we continued to pass immence herds of buffaloe all night as we had done in the latter part of the day."[18] On August 29, Clark wrote that from an "eminance I had a view of a greater number of buffalow than I had ever seen before at one time. I must have seen 20,000 of those animals feeding on this plain."[19]

There were numerous other eye-witness accounts of large herds of bison. In 1839, John Kirk Townsend, a young ornithologist who had accompanied Nathaniel Jarvis Wyeth on his second expedition to the Rocky Mountains and the Columbia River in 1824, published *Narrative of a Journey across the Rocky Mountains to the Columbia River.* He reported that while they were traveling along the Platte River in what is now Nebraska, they observed a huge herd of bison:

> Toward evening on rising a hill, we were suddenly greeted by a sight which seemed to astonish even the oldest amongst us. The whole plain, as far as the eye could discern, was covered by one enormous mass of buffalo. Our vision, at the very least computation, would certainly extend ten miles, and in the whole of this great space, including about eight miles in width from the bluffs to the river bank, there was apparently no vista in the incalculable multitude. It was truly a sight that would have excited even the dullest mind to enthusiasm.[20]

An even more dramatic account was provided by Richard Irving Dodge, who in May of 1871 traveled in a light horse-drawn wagon between two forts on the Arkansas River, a distance of about 34 miles. He reported, "At least 25 miles of this distance was through one immense herd, comprised of countless smaller herds of buffalo then on their journey north . . . ." He added, "The whole country appeared one great mass of buffalo, moving slowly to the northward; and it was only when actually among them that it could be ascertained that the apparently solid mass was an agglomeration of innumerable small herds, of from fifty to two hundred animals . . . ."[21] Based on information provided by Dodge, William Temple Hornaday, who conducted a study of American bison for the Smithsonian Institution, estimated the size of the herd Dodge saw to be 4,000,000, which, he added, "is more likely to be below the truth than above it."[22]

What Dodge witnessed was the last of the great herds. Thirteen years later there were only 325 wild bison left in the United States.[23] Their decline was as sudden and precipitous as that of the passenger pigeon.

What brought the American bison to near extinction? The story is much the same as that of the passenger pigeon. They were slaughtered by the millions by sportsmen who did not have a keen sense of what constituted sport, by commercial hunters who cashed in on the market for their hides and their tongues, and by hired gunmen whose job was simply to get rid of them.

While Native Americans killed bison to sustain their way of life, they did not kill so many that it disrupted the ecosystems of which they were a part and threatened the survival of species. That changed in the 1830s when European Americans began the mass destruction of what had once been huge herds of bison.[24]

The demise of once-thundering herds of bison on the Great Plains did not start with European-American hunters slaughtering them. That was to come later. Rather, it began with the commercialization of bison hunting as the American Fur Company and other trading companies bartered with members of various Native American tribes for bison hides[25] Spongier than leather made from the hides of cattle but also more elastic, leather made from the hides of bison was excellent for the belting needed by rapidly-expanding industries. Leather furniture and wall coverings, fashionable in the homes of the new industrial and financial barons, were made from the hides of bison. So also were the seats in carriages and sleighs. In the judgment of many, it was the best leather produced."[26]

In 1839, the American Fur Company purchased 45,000 bison hides. By 1847, the number of hides shipped to St. Louis had increased to 110,000.[27] Though no one realized it at the time, the destruction of the once-thundering herds of bison on the Great Plains was underway.

It also marked the beginning of the destruction of the way of life of the tribal groups that made the Great Plains their home. Fur companies, many of which were completely unscrupulous, supplied alcohol and credit to Native Americans to induce them to hunt extensively, killing as many bison as they could. The companies paid Native Americans rock-bottom prices for bison hides but charged them inflated prices for whatever they wished to purchase in return.[28]

Along with injecting alcohol in the veins of once-proud warriors, figuratively if not literally, fur companies changed the culture of the plains tribes in other ways as well. As noted, Native American hunters, with rare exception, only killed the number of bison that they needed to sustain their way of life. And they used just about everything that could be used from the bison they killed. The advent of the bison hide trade changed that. Spurred by a desire for alcohol and the goods trading companies offered at exorbitant prices, Native American hunters slaughtered large numbers of bison, taking only their hides and, in some cases, their tongues (for which there was also a market). Packs of wolves often followed the hunters at a safe distance to feast on the bison carcasses left behind.[29]

The European-American buffalo hunters who came to the Great Plains did not all come for the express purpose of slaughtering bison to supply the commercial market for hides and tongues, at least not initially. Some came to supply meat for the crews building the railways that were wending their way across the Great Plains, much as members of Lewis and Clark's expedition had done six decades earlier, albeit on a much smaller scale. Among those hired to provide meat for railroad crews was William F. ("Buffalo Bill") Cody, who in 1867–1868 worked for the Union Pacific Railroad to provide meat for the construction crews building what was to become part of the first transcontinental railroad.[30]

But in time, European-American commercial bison hunters flocked to the Great Plains in substantial numbers to slaughter bison for their hides and tongues. A well-equipped commercial hunting party consisted of one or two riflemen to shoot the bison, five or six skinners to remove the hides from the bison once they were killed, and two or three other men to stretch the hides on the ground at the hunting camp in order to dry them for shipment.[31] A favorite firearm of bison hunters was the Sharps Buffalo Rifle, which was manufactured in a variety of calibers though the .50 caliber was preferred by many hunters. A single-shot rifle with a long octagonal barrel, it was so heavy that most hunters fired it from a fixed position where they could rest the barrel on a rock or a tripod.[32] It was highly accurate at distances of 200 yards or more.[33]

Once the hunting parties had skinned the bison they had shot and, in some cases, cut out their tongues, the carcasses were left to rot. When the soft tissue had decomposed, scavengers collected the bones to ship back east. Some of the bones were used to make bone china.[34] Most of them, however, were ground to make fertilizer. In 1873, more than three million pounds of bison bones were shipped back east by one railroad company alone.[35]

In contrast to commercial hunters, who shot bison for their hides and tongues, those who viewed shooting bison as a sport shot them for whatever thrill they could get out of killing them. On one occasion when a large herd of bison crossed the tracks, stopping a train for two hours, many of those on the train amused themselves by shooting indiscriminately into the herd. When the herd had finally crossed the tracks and the train resumed its journey, there were more than 500 carcasses strewn about the prairie. All were left where they had dropped.[36]

One of the most notable hunts done in the name of sport occurred in 1872 when Grand Duke Alexis, the son of Emperor Alexander II of Russia, visited the United States. General Philip H. Sheridan, a Civil War hero who after the war was given command of all federal troops deployed on the Great Plains, was in charge of the hunt. The hunting party included several prominent members of the Russian nobility, senior Russian naval officers, and a large group of secretaries, valets, and servants. Lieutenant Colonel George A. Custer was among the Americans who accompanied the hunting party. A special train carried the hunting party and their supplies, which included a prodigious amount of champagne, to North Platte, Nebraska. There they were met by Buffalo Bill Cody, who had been hired to be a scout and an instructor for the Grand Duke. Cody allowed the Grand Duke to ride Buckskin Joe, his favorite horse for hunting bison,[37] and use his favorite hunting rifle, a single-shot .50 caliber Model 1866 Springfield which he had named "Lucretia Borgia."[38]

The Nebraska part of the hunt lasted two days. Estimates of the number of bison killed range from 500 to 1,500. The hunt did not end there, though. The special train took the hunting party to Colorado, where another two hundred bison were killed.[39]

General Sheridan participated in several other hunts organized for prominent people. One, which was held in 1884, included U.S. President Chester

A. Arthur, Secretary of War Robert Todd Lincoln (Abraham and Mary Todd Lincoln's eldest son), U.S. Senator George Graham Vest from Missouri, and Colonel John Schuyler Crosby, who was serving as the territorial governor of Montana. They had an escort provided by the Sixth U.S. Cavalry, forty pack mules, and a large number of servants and packers. The hunt was described as "the largest and most perfectly organized hunting party that ever slaughtered game wholesale for the pleasure it afforded."[40]

In 1870, two million bison were killed in Nebraska, Kansas, Texas and the Indian Territory (which subsequently became Oklahoma).[41] Each and every day, 5,000 bison were killed during 1872 and 1873. By 1876, the three to four million bison that had roamed the southern plains were gone.[42] (Lieutenant Colonel Custer also died that year as a result of some significant miscalculations that he made while commanding a detachment of the Seventh U.S. Cavalry at the Battle of the Little Bighorn River in Montana.)

When the southern herds of bison were gone, commercial hunters shifted their attention to the northern plains, where sizable herds remained. By the middle of 1883, virtually all of them were gone as well, with only a small remnant of the once-huge herds surviving. Many hunters thought the herds had moved north to Canada. That was not the case. The Canadian herds had been slaughtered as well.[43]

With the vast herds rapidly disappearing, some efforts were made to stop the slaughter, though these efforts encountered strong opposition. In testimony before Congress in 1874, Columbus Delano, a stern Ohioan who served as Secretary of the Interior during the presidency of Ulysses S. Grant, stated, "The buffalo are disappearing rapidly, but not faster than I desire. I regard the destruction of such game as Indians subsist upon as facilitating the policy of the Government, of destroying their hunting habits, coercing them on reservations, and compelling them to begin to adopt the habits of civilization."[44]

Similar sentiments were expressed by General Sheridan the following year in an address to the Texas state legislature. He suggested that every hunter should be given a medal with a dead buffalo on one side and a discouraged Indian on the other side. Echoing Secretary Delano, he added that once the bison were exterminated, the Indians would be controlled and civilization could advance.[45]

The mass slaughter of passenger pigeons and that of American bison occurred at roughly the same time. American bison might have suffered the same fate as passenger pigeons were it not for some timely and fortuitous intervention, some of it deliberate, some of it by happenstance.

YELLOWSTONE NATIONAL PARK

In 1872, Congress passed a bill that provided that "the tract of land in the Territories of Montana and Wyoming lying near the headwaters of the Yellowstone River . . . is hereby reserved and withdrawn from settlement, occupancy, or sale under the laws of the United States, and dedicated and set apart as a public park . . . for the benefit and enjoyment of the people . . . ." President Grant signed the act on March 1, 1872, which marked the birth of Yellowstone National Park, the nation's first national park.[46] Covering 2.2 million acres, the park includes rugged mountains, pristine lakes, rivers and streams, waterfalls, and geysers and hot springs sitting over a huge volcano waiting to erupt. It is also home to the only surviving herd of bison still living in the wild in the United States.

Though designating Yellowstone as a national park was supposed to protect the bison and other species living there, in practice this did not happen because of inadequate enforcement of the law. By 1884, there were only 25 bison left in the park.[47] In an effort to protect the remaining few, Congress gave the task of enforcing the law to the U.S. Army. The relevant legislation stated, "The Secretary of War, upon request of the Secretary of the Interior, is hereby authorized and directed to make the necessary details of troops to prevent trespassers or intruders from entering the park for the purpose of destroying the game or objects of curiosity therein, or for any other purpose prohibited by law, and to remove such persons from the park if found therein . . . ."[48] The National Park Protective Act (Lacy Act), enacted a decade later, was even more specific, stating that "all hunting, or the killing or wounding, or capturing at any time of any bird or wild animal, except dangerous animals, when it is necessary to prevent them from destroying human life or inflicting an injury, is prohibited . . . ."[49]

These measures protected the bison from further slaughter within the boundaries of the park. As a result, the Yellowstone herd has grown—slowly at first, more rapidly in recent years. Today the Yellowstone herd numbers more than 4,000. Because additional expansion of the herd would exceed the carrying capacity of the park, as well as cause problems when bison leave the park to graze on private land outside the park, the National Park Service, in cooperation with several other agencies, has taken measures to control the size of the herd.[50]

WALKING COYOTE

The success of the Yellowstone program is only part of the story of the rescue of bison from near extinction. Several private individuals—some intentionally, some inadvertently—played significant roles in the preservation of bison. Some were people of renown. Others were obscure individuals who are but footnotes in the annals of history. Walking Coyote was one of the latter.

A member of the North Pend d'Orielle tribe who lived in the Mission Valley in northwestern Montana, he was among those who played a role in the recovery of bison. He brought four orphaned bison calves—two heifers and two bulls—back with him when he returned from one of the last bison hunts on the east side of the mountains. When they were old enough to reproduce, Walking Coyote's little herd of bison grew in number.[51] After a few years, he sold the herd to Michel Pablo and Charles Allard, Sr., for $2,000.[52] Though no one realized it at the time, the portion of the herd he sold to Allard was to play a key role in efforts to save bison from extinction.

CHARLES E. CONRAD, SINGS-IN-THE-MIDDLE, AND THE CONRAD BISON HERD

Charles E. Conrad was another unlikely hero in the story of the rescue of bison from the brink of extinction. He was one of thirteen children born to Colonel James and Maria Conrad, who lived lives of luxury and privilege at Wapping, their plantation in the Shenandoah Valley of Virginia. The Civil War changed everything. During the last two years of the war, Charles and his older brother, William, rode with the 43rd Battalion of the Virginia Cavalry, which was commanded by John Singleton Mosby.[53] Often operating behind enemy lines, they struck swiftly and then disappeared, earning Mosby the nickname "The Gray Ghost."[54]

After the war, Charles and William, having lost everything, headed west to seek their fortune. They found it with I.G. Baker & Co., a transportation company that used riverboats and ox-drawn freight wagons to serve the regions that became Montana in the United States, and Alberta in Canada. A few years later, the brothers purchased I.G. Baker & Co. A considerable portion of the Conrad fortune was derived from trade in bison hides and tongues at a time when the herds were so immense that no one thought there was any risk of their becoming extinct.[55]

Charles Conrad came to identify strongly with Native American tribes, spending considerable time with various tribes while learning their languages and cultural traditions. In a Roman Catholic ceremony, he married Sings-in-the-Middle, the daughter of a Blackfoot chieftain who had converted to Roman Catholicism. When her father moved his tribe to Canada, she had to choose between going with her people or remaining with her husband and their young son, Charles E. Conrad, Jr. Sings-in-the-Middle chose the former, after extracting from her husband a promise to provide a Roman Catholic education for their son, a promise Charles kept.[56]

In time, the Conrad brothers sold I.G. Baker & Co. to Hudson's Bay Company and became involved in banking and real estate development. William founded the town of Conrad on the east side of Montana Rockies while Charles founded the town of Kalispell on the west side of the mountains, a town he named after the Kalispel tribe, a Salish tribe he held in high regard.[57]

In 1896, concerned about the diminishing number of bison, Conrad purchased fifty bison, which were part of the Allard herd.[58] After keeping them at various locations, Conrad moved his herd to a large fenced-in area on a hill north of Kalispell. The hill, now the site of a golf course, an extended care facility and a major medical center, continues to be known as "Buffalo Hill."[59]

After the death of Sings-in-the-Middle, Conrad married Alicia "Lettie" Davenport Stanford. Conrad died November 27, 1902.[60] His widow was to play a significant role in preserving the American bison.

THE AMERICAN BISON SOCIETY

On December 8, 1905, a group of sixteen conservationists met in the Lion House at the New York Zoological Park to organize the American Bison Society. Its mission was to protect the remaining herds of bison and encourage the establishment of new herds. William Temple Hornaday, whose book on the extermination of the American bison had established him as a leading authority on the subject, was elected president of the organization. President Theodore Roosevelt pledged his strong support. In a letter addressed to the secretary of the organization, he stated, "I feel real and great interest in the work being done by the American Bison Society to preserve the buffalo . . . . It would be a real misfortune to permit the species to become extinct, and I hope that all good citizens will aid the Society in its efforts for its preservation."[61] Roosevelt was named honorary president of the society.

Roosevelt made good on what he had promised. On May 23, 1908, he signed a bill authorizing the use of federal funds to purchase land suitable for the conservation of bison. It was the first time that Congress appropriated money from tax revenues to purchase land to conserve wildlife.[62] The land that was purchased for the National Bison Range is on rolling hills that overlook the Mission Valley in northwestern Montana. The Mission Mountains, majestic in their rugged beauty, rise above the east side of the valley.

The American Bison Society raised the money to buy the bison to be released on the range, once the fence was completed. They purchased thirty-four bison—twelve bulls and twenty-two cows—from the Conrad estate for $275 a head. Lettie Conrad donated two additional bison, selecting for donation the best two in the herd. They were delivered to the range on October 17,

1909. The American Bison Society then gave the bison to the federal government, which placed them under the care and protection of the Department of Agriculture.[63]

Today, between 350 and 500 bison roam the 18,800–acre National Bison Range.[64] In the course of the years, the federal government has donated a number of bison from the herd to other ranges and wildlife sanctuaries.

The American Bison Society played a key role in establishing other public herds as well. Then in 1935, confident that the future of the American bison was no longer at risk, the society voted itself out of existence.[65]

How many American bison are there today? Because many of them are in private herds, no one knows for certain. Defenders of Wildlife, a conservation group founded in 1947 dedicated to the protection of all native animals and plants,[66] estimates that there are approximately 500,000 bison in North America today. However, most of these are not pure wild bison, having been cross-bred with cattle and semi-domesticated.[67] Defenders of Wildlife estimates that there are fewer than 30,000 wild bison in conservation herds. Yellowstone National Park, with a bison population of approximately 4,000, has the largest herd of wild plains bison (*Bison bison bison*) while Wood Buffalo National Park in Canada, with a bison population of about 10,000, has the largest herd of wild wood bison (*Bison bison athabascae*).[68]

Though precise numbers with respect to the number of bison in North America today are difficult to pin down, what is certain is that they have been rescued from near extinction as a result of action taken by the federal government, private individuals, the American Bison Society, and other organizations and groups. Might the passenger pigeon also have been rescued had similar efforts been made to save it? We will never know the answer to this question.

NOTES

1. Defenders of Wildlife, "Basic Facts about Bison" at http://www.defenders.org/bison/basic-facts (last accessed June 10, 2019).

2. International Union for Conservation of Nature, "Bison bison," *Red List* at http://www.iucnredlist.org/details/2815/0 (last accessed June 10, 2019).

3. Some preferred clothing made from the hides of deer since it was softer than clothing made from the hides of bison.

4. Paul H. Carlson, *The Plains Indians* (Texas A&M University Press, 1998), 39–40.

5. Ben Pierce, "Places We Love: First Peoples Buffalo Jump State Park," *Bozeman Daily Chronicle*, April 16, 2015.

6. "All about Bison—Buffalo Jumps" at http://allaboutbison.com/buffalo-jumps/ (last accessed September 23, 2015).

7. Montana State Parks, "Madison Buffalo Jump State Park" at http://stateparks.mt.gov/madison-buffalo-jump/ (last accessed June 10, 2019). Because of financial problems, Madison Buffalo Jump State Park is in danger of being closed down. If that happens, the land could be sold to a developer for residential development (Bill Goold, "The Madison Buffalo Jump: Rich in Stories and Grandeur," *Distinctly Montana*, September 19, 2015, at http://www.

distinctlymontana.com/buffalo154#utm_sguid=89802,7678f8d7–d6f9–b973–5c8e-cc093b54ae3a (last accessed June 10, 2019).

8. Montana State Parks, "First Peoples Buffalo Jump State Park" at http://stateparks.mt.gov/first-peoples-buffalo-jump/ (last accessed June 10, 2019). A 2008 archeological survey commissioned by Montana Fish, Wildlife & Parks discovered projectile points (arrow heads and spear points) believed to be more than 5,000 years old, which suggest that First Peoples may have been one of the earliest bison jump sites on the Northern Plains. The survey also discovered tipi rings, 1,300 rock cairns, and 33 drive lines used to channel bison in the direction of the jump (Pierce, "Places We Love: First Peoples Buffalo Jump State Park.")

9. Erin Madison, "First Peoples Named National Historic Landmark," *Great Falls Tribune*, August 4, 2015.

10. Pierce, "Places We Love: First Peoples Buffalo Jump State Park."

11. *Original Journals of the Lewis and Clark Expedition 1804–1806,* Vol. II, Part 1(New York: Dodd, Meade & Company, 1904), 93. The spelling, capitalization, and punctuation are as they appeared in the original journal. When Lewis and Clark returned, there was no public recognition of what they had accomplished. Since they had failed to find an inland water route to the Pacific because none existed, President Jefferson, who had commissioned their epic voyage of discovery, showed little interest in their findings. Met with indifference, they took it upon themselves to publish their journals, which were republished in 1904. Only then did what they had accomplished attract the interest of the public.

12. Ibid.

13. United States Fish and Wildlife Service, "Time Line of the American Bison" at http://www.fws.gov/bisonrange/timeline.htm (last accessed June 10, 2019).

14. "Bison," *West Virginia (WV) Cyclopedia* at http://www.wvexp.com/index.php?title=Bison&oldid=27838 (last accessed June 10, 2019).

15. United States Fish and Wildlife Service, "Time Line of the American Bison."

16. *Original Journals of the Lewis and Clark Expedition 1804–1806,* Vol. V, Part 2, 199.

17. Ibid., 208.

18. Ibid., 226.

19. Ibid., 364.

20. John Kirk Townsend, "Across the Rockies to the Columbia" at http://user.xmission.com/~drudy/mtman/html/townsend/chap3.html (last accessed June 10, 2019).

21. Richard Irving Dodge, *The Plains of the Great West and Their Inhabitants* (New York: G. P. Putnam's Sons, 1876), 120.

22. Hornaday, *The Extermination of the American Bison*, 390–91

23. United States Fish and Wildlife Service, "Time Line of the American Bison."

24. United States Fish and Wildlife Service, "Time Line of the American Bison."

25. Andrew C. Isenberg, *The Destruction of the Bison: An Environmental History, 1750–1920* (Cambridge University Press, 2000), 105–106.

26. Mari Sandoz, *The Buffalo Hunters: The Story of the Hide Men* (New York: Hastings House Publishers, 1954), 96.

27. Isenberg, *The Destruction of the Bison*, 105–106.

28. Ibid., 106–107.

29. Ibid.

30. Joseph J. DiCerto, "William F. Cody," *Encyclopaedia Britannica Academic* at http://academic.eb.com./EBchecked/topic/124078/William-F-Cody (last accessed September 28, 2015).

31. Martin S. Garretson, *The American Bison: The Story of Its Extermination as a Wild Species and Its Restoration under Federal Protection* (New York Zoological Society, 1938), 115.

32. W. S. Peace, "Sharps Buffalo Rifle," *Handbook of Texas Online* at https://tshaonline.org/handbook/online/articles/lns02 (last accessed June 10, 2019). The price of a new Sharps .50 caliber rifle was usually around $100, the equivalent of more than $1,700 in today's dollars. Smaller calibers cost less (Garretson, *The American Bison*, 116).

33. Because of the Sharps's great accuracy when in the hands of a skilled marksman, it is believed to be the origin of the term "sharpshooter."

34. Compared to hard porcelain, bone china is somewhat easier to make and does not chip as readily as hard porcelain. The development of bone china has been attributed to Josiah Spode II, who introduced it around 1800. His basic formula, which is still in use today, called for six parts of bone ash, four parts of china stone, and three-and-a-half parts of china clay. Bone china has an ivory white appearance ("Bone China," *Encyclopaedia Britannica Academic* at http://academic.eb.com./EBchecked/topic/72917/bone-china?sections=72917main&cit=mla&view=print (last accessed September 28, 2015).

35. United States Fish and Wildlife Service, "Time Line of the American Bison."

36. Garretson, *The American Bison*, 112–113.

37. Garretson, *The American Bison*, 130–131.

38. Marguerita House, "Lucretia Goes Buffalo Hunting," Buffalo Bill Center of the West, 10 August 2014, at http://centerofthewest.org/2014/08/10/lucretia-hunts-buffalo/ (last accessed October 2, 2015). Lucretia Borgia (1480–1519) was the daughter of Pope Alexander VI, who, of course, wasn't supposed to have any children. Lucretia supposedly had a penchant for poisoning people, though those allegations have never been proven.

39. Garretson, *The American Bison*, 133–134.

40. Ibid., 136.

41. Sandoz, *The Buffalo Hunters*, 99.

42. United States Fish and Wildlife Service, "Time Line of the American Bison." See also Hornaday, *The Extermination of the American Bison*, 501–502.

43. United States Fish and Wildlife Service, "Time Line of the American Bison." See also Hornaday, *The Extermination of the American Bison*, 502–504, and Garretson, *The American Bison*, 151–155.

44. Quoted by Robert Wooster in *The Military and United States Indian Policy 1865–1903* (Yale University Press, 1988), 171.

45. United States Fish and Wildlife Service, "Time Line of the American Bison." It should be added that in later years, General Sheridan's views on the matter shifted somewhat. When General Alfred H. Terry in the late 1870s protested "the wholesale slaughter of buffalo on Indian reservations by white men," General Sheridan supported General Terry on this matter (Wooster, *The Military and United States Indian Policy 1865–1903*, 172).

46. *The Act of Dedication—March 1, 1872* (U.S. Code, Title 16, Chapter 1, Subchapter V) at http://www.yellowstone-online.com/history/yhfour.html (last accessed June 10, 2019).

47. United States Fish and Wildlife Service, "Time Line of the American Bison."

48. *Congressional Acts Pertaining to Yellowstone* at http://www.yellowstone-online.com/history/yhfour.html (last accessed June 10, 2019).

49. Ibid.

50. United States National Park Service, "Bison Management—Yellowstone National Park" at http://www.nps.gov/yell/learn/nature/bisonmgnt.htm (last accessed October 1, 2015).

51. James E. Murphy, *Half Interest in a Silver Dollar: The Saga of Charles E. Conrad* (Missoula, Montana: Mountain Press Publishing Company, 1983), 194–195.

52. Charlene Cross, *The Undying West: A Chronicle of Montana's Camas Prairie* (Golden, Colorado: Fulcrum Publishing, 1999), 173.

53. Candace Chase, "Kalispell Founder's Granddaughter Dies," *Daily Interlake*, August 6, 2010, at http://www.dailyinterlake.com/members/kalispell-founder-s-granddaughter-dies/article_5c14a7e0–a355–11df-bce5–001cc4c002e0.html (last accessed June 10, 2019).

54. Civil War Trust, "John Singleton Mosby" at http://www.civilwar.org/education/history/biographies/john-singleton-mosby.html?referrer=https://www.google.com/ (last accessed June 10, 2019).

55. Murphy, *Half Interest in a Silver Dollar*, 194.

56. Ibid., 43–44.

57. Murphy, *Half Interest in a Silver Dollar*, 144. See also "Kalispell," *Encyclopaedia Britannica Academic* at http://academic.eb.com./EBchecked/topic/681930/Kalispell (last accessed October 3, 2015).

58. United States Fish and Wildlife Service, "Time Line of the American Bison" and Chase, "Kalispell Founder's Granddaughter Dies." See also Murphy, *Half Interest in a Silver Dollar*, 195.

59. Franz, "Celebrating Conrad's Historic Herd." Before moving the herd to the land north of Kalispell, Conrad kept them on Wild Horse Island, a large island in Flathead Lake (the largest freshwater lake west of the Mississippi River) and on a ranch in Smith Valley, which is located west of Kalispell.

60. "Flathead County, Montana, Genealogy Trails."

61. Quoted by Garretson in *The American Bison*, 207.

62. United States Fish and Wildlife Service, "National Bison Range" at http://www.fws.gov/refuge/National_Bison_Range/about.html (last accessed June 10, 2019).

63. Garretson, *The American Bison*, 208–209.

64. United States Fish and Wildlife Service, "National Bison Range."

65. United States Fish and Wildlife Service, "Time Line of the American Bison." In 2005 the Wildlife Conservation Society relaunched the American Bison society. The reactivated American Bison Society has as its mission promoting larger, free-ranging herds (Wildlife Conservation Society, "The American Bison Society" at http://www.wcs.org/saving-wildlife/hoofed-mammals/bison/the-american-bison-society.aspx [last accessed June 10, 2019]).

66. Defenders of Wildlife, "Mission and Vision" at http://www.defenders.org/mission-and-vision (last accessed June 10, 2019).

67. The meat of bison that have been cross-bred with cattle (sometimes called "beefalo") is low in saturated fat and, accordingly, is quite healthy to consume. The growth of private herds of bison and beefalo has progressed to the point that buffalo meat is readily available for purchase. A number of restaurants, particularly in the western part of the United States and Canada, include buffalo burgers on their menus.

68. Defenders of Wildlife, "Basic Facts about Bison."

Chapter Three

Rescuing Species from the Brink of Extinction

As noted in the Prologue, Aldo Leopold states in *A Sand County Almanac*, "In short, a land ethic changes the role of *Homo sapiens* from conqueror of the land-community to plain member and citizen of it. It implies respect for his fellow-members, and also respect for the community as such."[1] The mass slaughter of passenger pigeons and bison were humankind at its worst—appalling mass destruction perpetrated by self-serving individuals who believed that they were conquerors of nature, which gave them license to do as they wished. In sharp contrast, Leopold insists that being a "plain member and citizen" of the land-community implies having respect for the other members and respect for the community as well. This singular insight provides the cornerstone for building an ethic that calls for protection of endangered species.

A similar theme is emphasized by Francis in *Laudato Si'*:

> It is not enough . . . to think of different species merely as potential "resources" to be exploited, while overlooking the fact that they have value in themselves. Each year sees the disappearance of thousands of plant and animal species which we will never know, which our children will never see, because they have been lost forever. The great majority become extinct for reasons related to human activity. Because of us, thousands of species will no longer give glory to God by their very existence, nor convey their message to us. We have no such right.
>
> It may well disturb us to learn of the extinction of mammals or birds, since they are more visible. But the good functioning of ecosystems also requires fungi, algae, worms, insects, reptiles and an innumerable variety of microorganisms.[2]

NATURAL EXTINCTIONS AND ANTHROPOGENIC EXTINCTIONS

As every school child knows, numerous species have become extinct in the last 500 million years or so, dinosaurs of different types being obvious examples. However, as Holmes Rolston, III, notes in *Environmental Ethics*, there is a huge difference between natural extinctions and anthropogenic extinctions caused by human encroachment. He observes:

> Relevant differences make the two as morally distinct as death by natural causes is from murder. Though harmful to a species, extinction in nature is no evil in the system; it is rather the key to tomorrow . . . . Such extinction is normal turnover in ongoing speciation.
>
> But anthropogenic extinction has nothing to do with evolutionary speciation . . . . In natural extinction, nature takes away life—when it has become unfit in habitat or when the habitat alters—and supplies other life in its place. Artificial extinction shuts down tomorrow because it shuts down speciation. Natural extinction typically occurs with transformation either of the extinct line or related or competing lines. Artificial extinction is without issue. One opens doors; the other closes them. In artificial extinction, humans generate and regenerate nothing; they only dead-end these lines.[3]

In short, when Martha, the last passenger pigeon, died, it was the end of a line. There were no new species that resulted from the extinction of passenger pigeons, no progeny—nothing at all. The risk of anthropogenic extinction of threatened and endangered species is where Rolston and other environmentalists appropriately focus their concern.

PRESERVING LINES OF LIFE

Rolston suggests, "[D]uties to a species are not to a class or category, not to an aggregation or average of sentient interests, but to a life line. An ethic about species needs to see how the species *is* a bigger event that the individual interests or sentience. Making this clearer can support a conviction that a species *ought* to continue."[4] He further observes:

> No individual crosses the extinction threshold; the species does. Reproduction is typically assumed to be a need of individuals, but since any particular individual can flourish somatically without reproducing at all . . . we can interpret reproduction as the species keeping up its own kind by reenacting itself again and again, individual after individual. It stays in place by its replacements.
>
> In this sense a female grizzly does not bear cubs to be healthy herself, any more than a woman needs children to be healthy. Rather, her cubs are *Ursus arctos*, threatened by nonbeing, recreating itself by continuous performance. A

> species in reproduction defends its own kind from other species . . . . The locus of the intrinsic value—the value that is really defended over generations—seems as much in the form of life, the species, as in the individuals . . . .[5]

It is for this reason that wildlife biologists, concerned about endangered species threatened with extinction, focus attention on the breeding population of that species, which, if it drops below the replacement level necessary for continuation of the species, places the species at risk for crossing the extinction threshold. It is also for this reason that there is a certain sense in which the extinction of *Ectopistes migratorius* occurred, not when Martha died in her cage at the Cincinnati Zoological Gardens, but several years earlier when the breeding population in the wild dropped below the critical mass necessary for continuation of the species.

INTRINSIC VALUE, INSTRUMENTAL VALUE, AND QUESTIONS OF PRIORITY

As noted in the Prologue, in an essay included in his book, *In Defense of the Land Ethic*, J. Baird Callicott concludes that as moral theories, "[the creation story found in Genesis 2] and bio-empathy appear to me to provide most effectively for the intrinsic value of other species,"[6]

While it might be possible to argue, at least if one uses the Genesis creation stories as reference points, that all species possess *intrinsic* value, it does not follow that all species possess the same *instrumental* value or *aesthetic* value. Take, for example, the *Anopheles* mosquito, which transmits malaria. Many would say that it not only lacks positive instrumental value but, though the term is a bit awkward, might even be said to have negative instrumental value—that is to say, be detrimental to those who contract malaria as a result of being bitten by an *Anopheles* mosquito. The notion of negative instrumental value might also be applied to black legged ticks (*Ixodes scapularis*), which spread Lyme disease, and several other species of plants and animals. Indeed, the negative instrumental value that these and other species have might well be sufficient in magnitude to offset any intrinsic value they might have and, in some instances, justify their eradication. For example, in order to reduce the risk of malaria being transmitted by *Anopheles* mosquitos, many public health authorities recommend spraying with insecticides swamps that are breeding grounds for these mosquitos.[7]

What is true of instrumental value is also true of aesthetic value, though this is somewhat more difficult to quantify. Suffice it to say that not everything that is part of the natural world is perceived to be aesthetically pleasing.

In short, when instrumental value and aesthetic value are taken into account, there is a case to be made for saying that all species do not have equal value. At the same time, as noted in the Prologue, if one assumes that all species have intrinsic value, it is plausible to argue that, particularly in the case of animals, all species ought to be treated with respect and dignity and not be tortured or otherwise be treated in an inhumane manner.

THE U.S. ENDANGERED SPECIES ACT OF 1973: A CASE STUDY

Because private property rights are strongly defended in the United States—to a greater degree than in most other countries—the controversy surrounding the U.S. Endangered Species Act of 1973 provides a useful case study of conflict between private property rights and protection of endangered species.

The vigorous defenders of liberty who framed the U.S. Constitution placed strong emphasis on the protection of property rights—protection that was made even more explicit in the Bill of Rights.[8] In an article published in the *National Gazette* on March 27, 1792, James Madison, the chief architect both of the U.S. Constitution and the U.S. Bill of Rights, commented, "Government is instituted to protect property of every sort . . . . This being the end of government, that alone is a *just* government, which *impartially* secures to every man, whatever is his *own*"[9]—a characterization of government reminiscent of John Locke's view that "the great and chief end . . . of men's uniting into commonwealths and putting themselves under government is the preservation of their property."[10] In short, the framers of the U.S. Constitution and Bill of Rights saw protection of property rights as essential if we are to experience liberty. For this reason, violations of property rights came to be viewed as violations of liberty.

While many in the United States today view environmental protection as the domain of the Democratic Party, it is interesting to note that the Endangered Species Act of 1973 became law during a Republican administration—that of Richard M. Nixon.[11] And it was enacted with broad bipartisan support, with the U.S. Senate voting 99–1 in favor of the bill and only twelve members of the U.S. House of Representatives voting against it.[12] Upon signing the law, President Nixon stated, "Nothing is more priceless and more worthy of preservation than the rich array of animal life with which our country has been blessed. It is a many-faceted treasure, of value to scholars, scientists and nature lovers alike and it forms a vital part of the heritage we share as Americans."[13]

While the act has been amended several times, most notably in 1978, 1982, 1988, and 2004, the basic provisions of the act have remained essentially unchanged.[14] The act defines:

- "endangered species" as "any species which is in danger of extinction throughout all or a significant portion of its range other than a species of

Class Insecta determined by the Secretary [of the Interior] to constitute a pest whose protection under the provisions of this Act would present an overwhelming and overriding risk to man" (Section 3[6]),

- "threatened species" as "any species which is likely to become an endangered species throughout all or a significant portion of its range" (Section 3[20]), and
- "critical habitat" as "the specific areas within the geographical area occupied by the species . . . on which are found those physical or biological features (I) essential to the conservation of the species and (II) which may require special management . . . ." The act, however, specifies that except in cases involving special circumstances critical habitat "shall not include the entire geographical area which can be occupied by the threatened or endangered species" (Section 3[5a]).[15]

The act mandates that that the Secretary of the Interior "shall by regulation . . . determine whether any species is an endangered species or a threatened species" and that "concurrently with making a determination" shall "designate any habitat of such species which is then considered to be critical habitat . . . . The Secretary shall designate critical habitat . . . on the basis of the best scientific data available and after taking into consideration the economic impact, and any other relevant impact . . . ."[16] The regulatory authority assigned to the Secretary of the Interior is delegated to the Fish and Wildlife Service while that assigned to the Secretary of Commerce is delegated to the National Marine Fisheries Service.

At first, there was little criticism of the Endangered Species Act. After all, who could be opposed to preserving bald eagles and other icons of the U.S. national heritage? But then matters started getting complicated.

The controversy about the Tellico Dam, a project of the Tennessee Valley Authority (TVA) on the Little Tennessee River, marked a turning point in public perception. In 1973, David Etnier of the University of Tennessee discovered a little fish called the snail darter (*Percina tanasi*) in a portion of the Little Tennessee River that would be affected by the dam. Lynn A. Greenwalt, the head of the Fish and Wildlife Service at the time, advised TVA to halt construction and encouraged them to look in nearby streams for other populations of the snail darter. TVA did neither. In October 1975, the Fish and Wildlife Service listed the little fish as an endangered species on an emergency basis and ruled that TVA could not complete construction of the dam. TVA refused to comply and the case ended up in court. The district court dismissed the case. An appeal was filed. The appellate court ruled that the Endangered Species Act did apply to Tellico and issued an injunction ordering TVA to halt construction on the project, which was 90 percent complete. TVA appealed the case to the U.S. Supreme Court.[17] The U.S. Supreme Court upheld the appellate court decision. Delivering the opinion of the Court, Chief Justice Warren Burger wrote, "It is clear from the Act's legislative history that Congress intended to halt and reverse the trend toward species extinction, whatever the cost." He further noted, "It may seem curious to some that the survival of a relatively small number of three-inch fish among all the countless millions of species extant would require the permanent halting of a virtually completed dam for which Congress has expended more than $100 million . . . . We conclude, however, that the explicit provisions of the Endangered Species Act require precisely that result."[18]

The Tellico Dam controversy, which was a clash between two agencies of the federal government, did not involve private property rights (though it did have some implications for private property owners hoping to benefit from the project). Yet, in a very direct and dramatic way, it illustrated the impact the Endangered Species Act could have. And it made it very clear that the act did not pertain only to bald eagles and other high-profile creatures that everyone wants to save but to obscure plants and animals as well.

Clashes involving private property rights were not far down the road. While the Fish and Wildlife Service initially focused its efforts on preserving critical habitat in national forests and on other publicly-held land, in the 1980s regulatory intervention was extended to private property as well.[19] When this happened, the Endangered Species Act, which had been almost universally applauded at its birth, became the center of heated controversy as good will gave way to anger and hostility.

In the 1990s, the Clinton administration placed restrictions on logging old-growth forest (including on some parcels that were privately owned) in order to protect the habitat of the northern spotted owl, a threatened species, and the red-cockaded woodpecker, an endangered species. A consortium of timber industry representatives and private property owners took legal action. In *Babbitt v. Sweet Home Chapter of Communities for a Great Oregon*, lawyers representing the consortium charged that the Secretary of the Interior had overstepped his authority by interpreting the Endangered Species Act's prohibition of harm to threatened and endangered species as including "significant habitat modification or degradation" that kills or injures threatened and endangered species by "significantly impairing essential behavioral patterns, including breeding, feeding or sheltering." The district court ruled against the petitioners, but the U.S. Court of Appeals for the District of Columbia reversed the lower court decision, with a majority of the court of appeals, upon rehearing the case, concluding that the Secretary of the Interior had exceeded the authority granted by the Endangered Species Act. The case was appealed to the U.S. Supreme Court. In a 6–3 decision, the Supreme Court reversed the court of appeals decision. The Court stated that the Endangered Species Act applies "to all land in the United States," not just land in the public domain. Thus, the Court gave its approval to regulatory intervention to prevent destruction of critical habitat on private property.[20] In practice, this means that if the Secretary of the Interior determines that there is habitat critical to the survival of endangered species on land owned by farmers, ranchers, and other private individuals, they are subject to legal penalties if they cut down trees or otherwise adversely affect this critical habitat.

TAKINGS?

The matter was not yet fully resolved. Private property rights advocates turned to the "takings" provision of the Fifth Amendment to claim a right to compensation for any diminished value to private property resulting from regulatory intervention limiting its use. The Fifth Amendment is widely (and appropriately) interpreted as requiring that the government compensate property owners for personal property that has been requisitioned and for land and other property taken for public uses such as building post offices, highways, and airports. But what if title to the property remains in the hands of the private property owner but the market value of the property diminishes because of laws that have been enacted or regulations the government has issued? Does this diminished value also constitute a taking for which the property owner is entitled to compensation? If a determination of critical habitat made by the Fish and Wildlife Service limits what a property owner can do with his or her property and, as a result, the market value of the property decreases, is the property owner entitled to compensation?

That was the issue at stake in *Good v. United States*, a case filed with the U.S. Court of Federal Claims in 1997. Lloyd A. Good, Jr., the owner of a 40–acre tract of land in the Florida Keys where he planned to develop a 54–lot subdivision with a 48–slip marina, alleged that the U.S. Army Corps of Engineers' refusal to grant him a permit to dredge and fill wetlands constituted a taking under the Fifth Amendment and entitled him to $2.5 million in compensation. Because the property provided habitat for the Lower Keys marsh rabbit and the silver rice rat, which were listed as endangered species, the Fish and Wildlife Service had recommended that the Corps of Engineers not grant the permit.[21]

The claims court ruled that no taking had occurred. Judge James F. Merow, who delivered the opinion of the court, noted that land development is "a highly regulated business, and plaintiff's sought uses for his property were subject to restriction or prohibition under this regulatory regime." Merow concluded, "While plaintiff was free to assume the investment risks involved after considering that regime, the Fifth Amendment does not require the federal government to act as his surety should that investment prove to be ill-taken."[22]

U.S. courts have tended to make a distinction between regulatory intervention that totally (or almost totally) destroys the value of property and regulatory intervention that merely diminishes the value of property. Take, for example, *Pennsylvania Coal v. Mahon*, a case decided by the U.S. Supreme Court in 1922. H.J. Mahon had purchased a home on land where the Pennsylvania Coal Company had retained ownership of subsurface coal deposits. Subsequent to McMahon's purchase of the home, Pennsylvania passed a law prohibiting the mining of coal under residences, thereby preventing Pennsylvania Coal from mining the coal. The U.S. Supreme Court ruled that this regulatory intervention constituted a taking. Justice Oliver Wendell Holmes, who delivered the opinion of the Court, stated:

> Government hardly could go on if to some extent values incident to property could not be diminished without paying for every such change in the general law . . . . But obviously the implied limitation must have its limits . . . . One fact for consideration in determining such limits is the extent of the diminution. When it reaches a certain magnitude, in most if not in all cases there must be an exercise of eminent domain and compensation to sustain the act.[23]

The U.S. Supreme Court issued a similar ruling in *Lucas v. South Carolina Coastal Council* a half century later. In 1977, David H. Lucas purchased two lots on a barrier island which, at the time of purchase, was zoned for single-family residential construction. He paid $975,000 for the two lots. In 1986, however, South Carolina enacted a beachfront management act that prohibited construction of permanent habitable structures on the island. The Court determined that the "takings clause of the Federal Constitution's Fifth Amendment is violated when land-use regulation . . . denies an owner economically viable use of his or her land."[24] A South Carolina court awarded Lucas $1,232,000 as compensation for the lost use of his lots.[25]

On the other hand, in *Miller v. Schoene*, the U.S. Supreme Court ruled that "the owner of ornamental cedar trees is not deprived of his property without due process of law by the state requiring the destruction of the trees to avoid the infecting of apple orchards in the vicinity with cedar rust."[26] And in *Penn Central Transportation Co. v. New York City* (438 U.S. 104 [1978]), the U.S. Supreme Court ruled that New York City's Landmarks Preservation Law, which limited what Penn Central could do with the airspace above Grand Central Terminal, "did not effect a 'taking' of private property by the government without just compensation in violation of the Fifth and Fourteenth Amendments, since (1) the law did not interfere with the present uses of the building . . . , (2) the law did not necessarily prohibit occupancy of any of the air space above the landmark building, since under the procedures of the law, it was possible that some construction in the air space might be allowed, and (3) the law did not deny all use of the owner's preexisting air rights above the landmark building . . . ."[27]

THE ETHICS OF REGULATORY INTERVENTION

Regulatory intervention, which by its very nature often interferes with liberty and individual rights such as property rights, is often costly. But just as war is sometimes the lesser of evils, so also is regulatory intervention, including that directed toward protecting endangered species.

Traditional just war theorists have always insisted that a just cause, while a necessary condition, is not a sufficient reason to justify military action. Other criteria must also be satisfied, among them last resort, lawful authority, reasonable hope of success, and due proportionality.[28] Just war theorists have consistently argued that military action should always be a last resort taken only after other ways of addressing the problem have been exhausted. The same is true of regulatory intervention. That is why fostering civic virtue and soft power approaches are the place to begin, not regulatory intervention. Moreover, all regulatory intervention must be done with due diligence to lawful authority and to due process which protects the rights of individuals. And as with military action, regulatory intervention is justifiable only if there is reason to believe that it will be successful in resolving the problem that is being addressed. Finally, regulatory intervention is justifiable only if the marginal benefit of the intervention is sufficient to offset the costs related to it—costs which include both monetary costs and costs related to restrictions of individual liberty. Each of these criteria—just cause, last resort, lawful authority, reasonable hope of success, and due proportionality—merits more detailed comments.

1. *Just Cause.* Any intervention in the lives of individuals, be it in the form of environmental regulations or in any other way, is justified only if it contributes in a significant way to the common good;
2. *Last Resort.* Coercive intervention, such as protecting wetlands on private property, should be done only as a last resort after non-coercive approaches, such as providing financial incentives, have failed;
3. *Lawful Authority.* Any coercive intervention directed toward ensuring the common good must be done with diligent attention to due process and established law and to the rights of all individuals affected by the action in question;
4. *Reasonable Hope of Success.* Because intervention is always costly, it should be done only if there is good reason to believe that it will be successful in accomplishing its objectives; and
5. *Due Proportionality.* The good to be realized must be sufficient to offset the costs of the intervention, including both the cost to individual liberty and the administrative and other economic costs of the intervention itself. U.S. Supreme Court Justice Antonin Scalia, who delivered the opinion of the Court in *Michigan v. Environmental Protec-*

tion Agency, got it right when he stated that the U.S. Environmental Protection Agency "must consider cost—including most importantly, cost of compliance—before deciding whether regulation is appropriate and necessary."[29]

THE RATIONALE FOR PROTECTING ENDANGERED SPECIES

President Nixon got it right when, as noted earlier in this chapter, he stated with respect to the Endangered Species Act of 1973, "Nothing is more priceless and more worthy of preservation than the rich array of animal life with which our country has been blessed. It is a many-faceted treasure, of value to scholars, scientists and nature lovers alike and it forms a vital part of the heritage we share as Americans." It should be added that if, as has been suggested with respect to the creation stories in Genesis, there is an intrinsic goodness in all of creation, the rich array of species to which President Nixon referred needs to be expanded to include plants, as well as animals—a view consistent with Leopold's notion of the land community and Francis's observations in *Laudato Si'*.

At the same time, as was suggested above, when instrumental value (including value to species other than *Homo sapiens*) and aesthetic value are added to the equation, all species are not of equal value, a basic fact of life that is not always recognized by advocates of protecting endangered species. This means that there is a case to be made for establishing priorities with respect to protection of endangered species. And, as was noted earlier in this chapter, there is a case to be made for giving priority to protecting species threatened with anthropogenic extinction as a result of human encroachment.

LAST RESORT

The criterion of last resort is exceedingly important in view of the fact that those who view the world from the perspective of regulatory bureaucracies are often quick to exercise the power of regulation placed in their hands without exploring the possibilities for resolving the particular issues in other ways. Sometimes there are other options—options which ought to be explored rather than too readily curtailing the rights of private property owners and others by imposing mandatory solutions.

The example of the greater sage grouse (*Centrocercus urophasianus*) illustrates the way that cooperative approaches to protecting species at risk can ensure their survival without resorting to mandated solutions limiting what private property owners can do with their property.

Sage grouse are showy birds. The males have spiked tail feathers and two yellow sacs on their necks that they inflate when trying to impress females.

Sage grouse, which eat only sage brush during the winter,[30] once numbered in the millions. Because of loss of habitat,[31] the population has diminished to somewhere between 250,000 and 500,000 birds, spread out over 165 million acres in eleven states in the American West. The federal government owns more than 100 million of those acres, with much of the remainder in private hands.[32]

Concern about the declining population led to a debate extending over several years as to whether the greater sage grouse should be listed as an endangered species, which would have brought into play restrictions intended to preserve critical habitat. Opponents of listing the bird as an endangered species cited a study which indicated that listing the sage grouse could result in an annual loss of $5.6 billion in economic output.[33] Conservationists insisted that protecting the sage grouse is essential if it is to be saved from extinction.

After much discussion and extensive dialogue with interested parties, Secretary of the Interior Sally Jewell announced on September 22, 2015, that the department was not going to list the sage grouse as an endangered species. Instead, the federal government, state governments, ranchers, energy companies, and others agreed to work together to preserve habitat essential for survival of the sage grouse.[34] Audubon Society Vice President Brian Rutledge characterized the joint effort as "unprecedented." Commenting on the agreement, he stated, "We have gotten attention paid to an ecosystem that was at best forgotten and at worst being destroyed by the way it was being managed, and we are seeing that turned about."[35] Paul Ulrich, the regulatory director at Jonah Energy, an oil and gas company, commented, "Today's decision validates that we can work together to find solutions. We've already seen a shift by industry as our knowledge about the greater sage grouse has improved."[36] It was a win-win situation for all involved—and perhaps most importantly, for the greater sage grouse (though whether that continues to be the case remains to be seen as the Trump administration considers scaling back protection for the greater sage grouse).[37]

The Department of the Interior is not the only branch of the U.S. federal government that has worked cooperatively with state governments and the private sector to protect the sage grouse. The Natural Resource Conservation Service (NRCS), which is part of the U.S. Department of Agriculture, took the lead role in establishing the Sage Grouse Initiative (SGI). The NRCS website describes it as follows: "The Sage Grouse Initiative (SGI) is a partnership of ranchers, agencies, universities, non-profit groups, and businesses that embrace a common vision: wildlife conservation through sustainable ranching."[38] Noting that in states such as Montana a significant portion of the sage grouse's habitat is on private land, NRCS Chief Jason Weller observes that the initiative "is living proof that wildlife and agriculture can coexist and thrive in harmony." He adds, "SGI is making a difference because private landowners voluntarily work with us to produce results on the ground."[39] Millions of acres of private land stretching across several Western states are now protected by conservation easements.[40] That is good news for sage grouse and for numerous other species as well. In short, there often are alternatives to regulatory intervention that can protect endangered species. These alternatives both can and should be explored first, rather than too readily opting for regulatory intervention. At the same time, one should not assume that endangered species in all cases can adequately be protected without regulatory intervention. It should be the last resort, however, not the first resort.[41]

DILIGENT ATTENTION TO DUE PROCESS AND TO THE RIGHTS OF ALL INDIVIDUALS

There is no reason to doubt that defenders of the Endangered Species Act have anything other than the best of intentions when advocating regulatory intervention to protect threatened and endangered species. And since the Endangered Species Act has withstood the test of constitutional review, the matter of lawful authority has been resolved to the satisfaction of all but the most diehard critics of the act. That, however, does not completely resolve the matter. How authorities go about exercising the power entrusted to them by the provisions of the Endangered Species Act makes a huge amount of difference. If authorities are respectful and considerate of those who are subjected to the provisions of the act, much of the sting of being subjected to the exercise of this power is mitigated. On the other hand, if those who wield the power of regulatory intervention are inebriated by the power they possess and use it without regard to the feelings and sensitivities of those subjected to it, the human costs of regulatory intervention are exacerbated greatly.

REASONABLE HOPE OF SUCCESS

If regulatory intervention is to be morally justifiable, there must be reason to believe that intervention can protect endangered species from extinction. Assessment of the prospects for success of particular protection programs is best left to wildlife biologists, agronomists, and other scientists. Note should be taken, however, of successful efforts to protect endangered species:

- In 1985, the last nine remaining California condors were taken into captivity to begin captive breeding programs at the San Diego and Los Angeles zoos. In 1991 they were reintroduced into the wild in southern California and five years later in Northern Arizona. In 2004, California condors reproduced in the wild for the first time in seventeen years.

- In 1994, the Arctic peregrine falcon, which had been on the endangered species list, was delisted due to recovery.
- In 1999, the American peregrine falcon was delisted due to recovery.
- In 2001, the Aleutian Canada goose was delisted due to recovery.
- In 2007, the bald eagle was delisted due to recovery.[42] The bald eagle population, which had dropped to as few as 500 nesting pairs in the lower forty-eight states, has rebounded substantially. Today there are more than 5,000 nesting pairs in the lower forty-eight states. In the United States and Canada combined, there are now somewhere around 70,000 bald eagles.[43]
- Once exterminated to the point of near extinction, gray wolves, as a result of protection extended by the Endangered Species Act and the efforts of conservationists, have experienced a remarkable recovery in two parts of their former range—the western Great Lakes region and in the northern Rockies.[44]
- Grizzly bears, which, like gray wolves, were once hunted and exterminated to the point of near extinction, have rebounded in two of the five ecosystems in the lower forty-eight states in which they are found—the Yellowstone Ecosystem and the Northern Continental Divide Ecosystem.[45]

While the Endangered Species Act has played a significant role in the recovery of these species, this is not the sole reason that the populations of these and other endangered species have rebounded. Other factors include the work done by the Audubon Society, Defenders of Wildlife, and other conservation organizations, and by private individuals who have placed conservation easements on their property to preserve critical habitat. Yet another factor is the action the U.S. Environmental Protection Agency (EPA) took in 1972 banning the use of DDT (dichlorodiphenyltrichloroethane), a potent pesticide that resulted in a good deal of collateral damage such as thinning the eggshells of bald eagles and peregrine falcons, adversely affecting their prospects for hatching.[46] In various combinations, all of these factors have helped create the conditions which have enabled threatened and endangered species to rebound.

PROPORTIONALITY

If regulatory intervention is to be justified, the good to be realized must be sufficient to offset the costs of the intervention, including both the cost to individual liberty and the administrative and other costs of the intervention itself. In an era in which a significant number of farmers, ranchers, and other property owners help preserve endangered species by placing conservation easements on their land and by otherwise participating in habitat conservation efforts, a key issue here is whether the marginal benefits of regulatory intervention in a particular case would be sufficient to offset the costs related to it. Take, for example, the sage grouse initiative (SGI). Suppose for the sake of the discussion that listing the sage grouse as an endangered species, which would bring into play legal restrictions on what property owners could do with their property if it was categorized as critical habitat, could maintain the sage grouse population at, say, a 5 percent higher level than what might be accomplished by SGI alone. As long as what might be accomplished by SGI alone ensures the continuation of a viable sage grouse population, thereby protecting it from extinction, would this extra 5 percent be sufficient to justify curtailing the rights of private property owners by limiting what they can do with their property? That would be a hard sell to those who place strong emphasis on private property rights.

Moreover, regulatory intervention might even be counter-productive in some cases. In a study of the Endangered Species Act published a few years ago, David S. Wilcove, Michael J. Bean, Robert Bonnie, and Margaret McMillan observe:

> Many landowners are capable of helping endangered species by creating, restoring, or enhancing habitat on their land, but are unwilling to do so. Their unwillingness often stems from fear of new restrictions. They are afraid that if they take actions that attract new endangered species to their land or increase the populations of the endangered species that are already there, their "reward" for doing so will be more regulatory restrictions on the use of their property. In its most extreme manifestation, this fear has prompted some landowners to destroy unoccupied habitats of endangered species before the animals could find it. One landowner, referring to the presence of red-cockaded woodpeckers on a small section of his property, announced, "I cannot afford to let those woodpeckers take over the rest of my property. I'm going to start massive clearcutting."[47]

Even when there are strong arguments to be made for preserving endangered species such as gray wolves and grizzly bears, there are often counter-arguments made by ranchers and farmers who believe that their livestock might be killed by predators such as wolves and grizzly bears—and with some evidence to substantiate that concern. In an article entitled "For Gray Wolves, A Success Story not Without Distractors" that ran in the *New York Times*, Clyde Haberman notes, "Even though, over all, [ranchers'] livestock herds held fairly steady, ranchers in some locations were badly hurt as the wolf packs multiplied. One man, for example awoke to find that 120 of his sheep had been killed overnight. Some who raised cattle fretted about calves that disappeared or injured themselves trying to escape the predators."[48]

Several private organizations and government agencies have programs to ameliorate these losses by providing compensation to ranchers and farmers who lose livestock to protected predators (which is referred to in the literature as "depredation").[49] Over a period of 23 years, Defenders of Wildlife, which spearheaded efforts to restore wolves to Yellowstone Park and other areas, paid more than $1.4 million to livestock owners for verified losses to wolves. With many states now providing compensation for livestock lost to predators, Defenders of Wildlife has shifted its funding to other areas.[50] States that provide compensation to livestock owners for livestock lost to predators include Minnesota,[51] Oregon,[52] Washington,[53] and Montana.[54]

The bottom line in all of this is that there are often collateral costs related to protection of endangered species which need to be taken into account. The Endangered Species Act of 1973, it is worth reiterating, states, "The Secretary [of the Interior] shall designate critical habitat . . . on the basis of the best scientific data available and after taking into consideration the economic impact, the impact on national security, and any other relevant impact . . . ."[55] In short, the act was never intended to be one-dimensional but rather calls for taking into consideration, not only impact on endangered species, but also economic impact and other factors. It implicitly gives recognition to the economic tether to which Leopold made reference. Even if one believes, as I do, that gray wolves and grizzly bears ought to be protected, the collateral costs must be taken into consideration and ameliorated whenever possible.

SUMMING IT ALL UP

The senseless slaughter of passenger pigeons and American bison are travesties that are huge stains on the American character—travesties that resulted in the extinction of passenger pigeons and the near extinction of American bison. As noted in the previous chapter, however, the story of American bison is also the story of redemption, a story of steps taken by private individuals such as Charles E. Conrad, nongovernmental organizations (NGOs) such as the American Bison Society and various branches of government to ensure that the American bison did not suffer the same fate as the passenger pigeon.

The passage of the U.S. Endangered Species Act of 1973 marked the beginning of a new chapter in the struggle to protect endangered species. Like the sustainable agriculture and recycling movements, this endeavor has achieved considerable success. But like the sustainable agriculture and recycling movements, much remains to be done, not just in the United States but in Africa and other parts of the world as well, with a huge number of species on the cusp of extinction,[56] much of it the result of human impact on the ecological landscape.[57]

When the U.S. Supreme Court ruled in *Babbitt v. Sweet Home* that the Endangered Species Act applies "to all land in the United States," not just land in the public domain, the protection of endangered species came into conflict with private property rights. Ideally, as is illustrated by the SGI, government agencies ought to work cooperatively with farmers, ranchers, and other property owners, as well as business interests, to protect endangered species while recognizing the economic interests of farmers, ranchers, and others. Whenever possible, government ideally ought to function as a facilitator, rather than as a regulator. In some situations, though, this might not be feasible, which means that in a certain range of cases, regulatory intervention might be a necessary last resort after, as the Endangered Species Act specifies, economic impact and other relevant factors are taken into consideration.

NOTES

1. Aldo Leopold, *A Sand County Almanac with Essays on Conservation from Round River* (New York: Ballantine Books, 1966), 240.
2. Francis, *Laudato Si': On Care for Our Common Home*, May 24, 2015, paragraphs 33–34, at http://w2.vatican.va/content/francesco/en/encyclicals/documents/papa-francesco_20150524_enciclica-laudato-si.html (last accessed April 30, 2019)
3. Holmes Rolston, III, *Environmental Ethics: Duties to and Values in the Natural World* (Philadelphia: Temple University Press, 1988), 154–155.
4. Ibid., 147.
5. Ibid., 148.
6. J. Baird Callicott, *In Defense of the Land Ethic: Essays in Environmental Philosophy* (State University of New York Press, 1989), 154. While Callicott's arguments are plausible, bio-empathy might not provide a rationale for viewing all species as having intrinsic value. For example, humans are arguably less likely to have bio-empathy for centipedes and nematodes than for deer and antelope.
7. DDT (dichlorodiphenyltrichloroethane), a potent insecticide that at one time was widely used in the United States and other countries, is now banned in the United States and many other countries but is still used India and some other countries to control malaria by eliminating *Anopheles* mosquitos. DDT decomposes slowly and has significant residual environmental costs. *The Stockholm Convention on Persistent Organic Pollutants*, an international treaty that went into effect in 2004, calls upon signatory nations to end agricultural use of DDT but grants an exception for disease control.
8. The United States Constitution, as originally drafted and ratified, says relatively little about property rights. Alexander Hamilton argued that this was not really a problem because the power of the federal government was limited to what the Constitution explicitly authorized it to do. Hamilton wrote, "For why declare that things shall not be done that there is no power to do? Why, for instance, should it be said that liberty of the press shall not be restrained, when no power is given by which restriction be imposed?" (Alexander Hamilton, "The Lack of a Bill of Rights," *Federalist No. 84* in Alexander Hamilton, James Madison and John Jay, *The Federalist*, ed. Benjamin Fletcher Wright [Cambridge, MA: The Belknap Press of Harvard University Press, 1961], 531). However, those concerned about individual rights, including property rights, wanted a more explicit prohibition of encroachment on individual liberties. The result was the Bill of Rights, which, as Bernard H. Siegan notes, "shielded from federal intrusion those property interests of most concern during that period: one's home, land, office, firearms, and financial resources" (Bernard H. Siegan, *Property and Freedom: The Constitution, the Courts, and Land-Use Regulation* [New Brunswick: Transaction Publishers, 1997], 20).

9. James Madison, "Property," *National Gazette*, March 27, 1792, in *The Papers of James Madison*, Vol. 14, ed. Robert A. Rutland *et al.* (University of Virginia Press, 1983), 266.

10. John Locke, *The Second Treatise of Government*, ed. Thomas P. Peardon, Library of Liberal Arts Edition (Indianapolis: Bobbs-Merrill Company, Inc., 1952), 71 (IX, 124).

11. The Endangered Species Preservation Act of 1966 and the Endangered Species Conservation Act of 1969 helped set the stage for the Endangered Species Act of 1973. A conference held in Washington, D.C. in 1973 resulted in the Convention on International Trade in Endangered Species of Wild Fauna and Flora (CITES), which restricted international commerce in plant and animal species believed to be threatened. The Endangered Species Act of 1973, passed later that year, combined and substantially strengthened the provisions of previous legislation, while also breaking new ground (U.S. Fish & Wildlife Service, "A History of the Endangered Species Act of 1973" at http://www.fws.gov/endangered/laws-policies/esa-history.html [last accessed June 10, 2019]).

12. When the U.S. House of Representatives originally voted on the bill, the vote was 390–12 in favor. When the final vote was taken after the bill came out of conference committee, the vote was 355–4 in favor. The original vote in the Senate was 99–1 in favor with the bill gaining approval without dissent after it came out of conference committee (*Congressional Record*, Vol. 119, Part 33 [December 19 and 20, 1973], 42528–42535, 42910, 42915–41916).

13. Quoted by Bonnie B. Burgess in *Fate of the Wild: The Endangered Species Act and the Future of Biodiversity* (Athens, Georgia: The University of Georgia Press, 2001), 3. The signing of the act, it is interesting to note, did not make much of a ripple. The *New York Times*, for example, buried the report of the signing of the bill as a brief addendum to a story about another bill the president signed the same day ("President Signs Manpower Bill: States and Communities Get Larger Role in Job and Training Programs," *New York Times*, December 28, 1973, 1, 13). The U.S. Environmental Protection Agency, which opened for business on December 2, 1970, was also a product of the Nixon presidency (Jack Lewis, "The Birth of EPA," *EPA Journal*, November 1985, at http://www2.epa.gov/aboutepa/birth-epa [last accessed October 12, 2015]).

14. "History of the Endangered Species Act of 1973."

15. Endangered Species Act of 1973 at http://www.fws.gov/endangered/laws-policies/section-3.html (last accessed June 10, 2019). And what constitutes a species? The Act states, "The term 'species' includes any subspecies of fish or wildlife or plants, and any distinct population segment of any species or vertebrate fish or wildlife which interbreeds when mature" (Sec. 3 [16]).

16. Endangered Species Act of 1973, Sec. 4 (1), (3) at http://www.fws.gov/endangered/laws-policies/section-4.html (last accessed June 10, 2019).

17. Burgess, *Fate of the Wild*, 10–11.

18. Supreme Court of the United States, *Tennessee Valley Authority v. Hill et al.*, 437 U.S. 153, 172–73 (1978). The following year, however, Sen. Howard Baker and Rep. John Duncan, Sr., both of Tennessee, added an amendment to the Energy and Water Development Appropriations Act exempting the Tellico project from the Endangered Species Act—legislation that became law when the act was signed by President Jimmy Carter. Subsequently, several other populations of the snail darter were discovered, and it was listed as a threatened species, rather than as an endangered species (U.S. Forest Service History, "1979: Snail Darter Exemption Case" at http://www.foresthistory.org/ASPNET/Policy/northern_spotted_owl/1979owl.snaildarter.aspx [last accessed October 12, 2015]).

19. During this same period, the number of threatened and endangered species listed by the Fish and Wildlife Service increased substantially, which had the practical impact of increasing the number of cases in which intervention was a possibility.

20. Supreme Court of the United States, *Babbitt v. Sweet Home Chapter of Communities for a Great Oregon,* 515 U.S. 687, 698 (1995).

21. U.S. Court of Federal Claims, *Good v. United States*, 39 Fed.Cl. 81 (1997). Though in an earlier opinion the Fish and Wildlife Service had concluded that the proposed project would have no effect on the endangered rabbit, the agency, after further study, noted that the populations of both the marsh rabbit and the silver rice rat were declining and concluded that develop-

ment of the property would place the continued existence of these endangered species in jeopardy.

22. *Good v. United States* (1997).

23. *Pennsylvania Coal v. Mahon*, 260 U.S. 393 (1922).

24. *Lucas v. South Carolina Coastal Council*, 505 U.S. 1103 (1992).

25. Bernard H. Siegan, *Property and Freedom: The Constitution, the Courts, and Land-Use Regulation* (New Brunswick: Transaction Publishers, 1997), 128.

26. *Miller v. Schoene*, 276 U.S. 272 (1928).

27. *Penn Central Transportation Co. v. New York City*, 438 U.S. 104 (1978).

28. Just war theorists do not always agree about which criteria should be used, though they do agree that all relevant criteria must be satisfied before resorting to military action is morally justifiable. In *Contra Faustum*, Augustine states with respect to the question of what constitutes a just war, "A great deal depends on the causes for which men undertake wars, and on the authority they have for doing so; for the natural order which seeks the peace of mankind, ordains that the monarch should have the power of undertaking war if he thinks it advisable, and that the soldiers should perform their military duties in behalf of the peace and safety of the community" (*Contra Faustum*, XXII, 75, tr. Richard Stothert, in *A Select Library of the Nicene and Post-Nicene Fathers of The Christian Church*, Vol. IV, *Writings in Connection with the Manichaean Controversy,* ed. Philip Schaff [Grand Rapids, MI: Wm. B. Eerdmans Publishing Company, 1887], 301). In *Summa Theologica*, Thomas Aquinas suggests that three conditions must be met if a war is to be just: "First, the authority of the sovereign by whose command the war is to be waged. For it is not the business of a private person to declare war . . . . Second, a just cause is required, namely that those who are attacked should be attacked because they deserve it on account of some fault . . . . Third, it is necessary that the belligerents should have a right intention, so that they intend the advancement of good, or the avoidance of evil" (*Summa Theologica*, II–II, Q.40, A. 1, tr. Fathers of the English Dominican Province, rev. Daniel J. Sullivan, *Great Books of the Western World, Vol XX* [Chicago: Encyclopedia Britannica, Inc., 1952], 578). Hugo Grotius, while deemphasizing *jus ad bellum* (justice in going to war) and placing greater emphasis on *jus in bello* (justice in the conduct of war) insisted that war be conducted under the authority of a sovereign and that a declaration of war be issued: "*It is essential to the nature of a public war that it should have the support of the sovereign power . . . . A declaration of war is also requisite*" (*The Law of War and Peace [De Jure Belli ac Pacis Libri Tres]*, III, iv–v, tr. Francis W. Kelsey [Indianapolis: The Bobbs-Merrill Company, Inc., 1925], 633). Summarizing classical just war theory, James Turner Johnson lists the major *jus ad bellum* criteria as just cause, right authority, right intent, proportionality, the end of peace and last resort (*Can Modern War Be Just?*, 18). James F. Childress lists the traditional *jus ad bellum* criteria as just cause, last resort, announcement of intentions, reasonable hope of success, proportionality and just intention (James F. Childress, "Just-War Criteria" in *War or Peace? The Search for New Answers*, ed. Thomas A. Shannon [Maryknoll, NY: Orbis Books, 1980], 46–48), while Ralph Potter adds highest lawful authority to the list (*War and Moral Discourse* [Richmond, VA: John Knox Press, 1969], 43–44).

29. *Michigan v. Environmental Protection Agency,* 576 U.S. (2015). The EPA had concluded that "costs should not be considered" when determining whether emissions from power plants should be regulated.

30. Sage brush is a generic term for a woody shrub that grows in arid areas of the American west. The most common type is *Artimesia tridenta.*

31. The loss of habitat is not entirely the result of oil and gas drilling and other forms of economic activity. Wildfires have destroyed the vegetation on numerous tracts of land that once provided habitat for sage grouse. Cheat grass (*Bromus tectorum*), a non-native invasive species of grass, often springs up in areas stripped bare by wildfires, preventing the regeneration of sage brush.

32. Martha Harbison, "What's up with the Greater Sage Grouse? A Definitive Guide to the Most Talked-About Bird in the West," *Audubon Newsletter*, 29 May 2015 at https://www.audubon.org/news/whats-greater-sage-grouse (last accessed June 10, 2019). Some of the sage brush habitat is owned by the eleven states in which there are populations of sage grouse.

33. "Economic Impact of 2013 BM Sage Grouse Conservation Plan," 1 March 2014, at http://www.westernenergyalliance.org/sites/default/files/Sage%20Grouse%20Economic%20Report%20–%20Final%20from%20Minuteman%20Press.pdf (last accessed June 10, 2019).

34. The decision was made somewhat easier by the fact that a Western Association of Fish and Wildlife Agencies (WAFWA) report released 17 August 2015 indicates that the greater sage grouse population is rebounding in some areas (Western Association of Fish and Wildlife Agencies, "WAFWA Report Documents Greater Sage-Grouse Population Rebound," August 17, 2015, at http://www.wafwa.org/news/e_1606/News/2015/8/_WAFWA_Report_Documents_Greater_Sage-Grouse_Population_Rebound.htm [last accessed June 10, 2019]).

35. Quoted by Clifford Krauss in "U.S. Trying to Protect Sage Grouse Without Listing It as an Endangered Species," *New York Times*, September 22, 2015 at http://www.nytimes.com/2015/09/23/business/us-trying-to-protect-sage-grouse-without-listing-it-as-an-endangered-species.html?_r=0 (last accessed June 10, 2019).

36. Quoted in ibid.

37. The Trump administration announced plans to open nine million acres of public land to drilling and mining by stripping away protections for the sage grouse (Coral Davenport, "Trump Drilling Plan Threatens 9 Million Acres of Sage Grouse Habitat," *New York Times*, December 6, 2018). Interior Secretary Ryan Zinke claimed this would not harm the sage grouse. "No one loves the sage grouse more than I do," he has stated (Quoted in ibid.). Environmentalists question that claim. Bobby McEnaney, who is affiliated with the Natural Resources Defense Council observes, "It's hard to pretend at this point that Zinke is a steward of America's public lands." McEnaney further notes, "With this single action, the administration is saying: This landscape doesn't matter. This species doesn't matter. Oil and gas matter" (Quoted in ibid.).

38. USDA Natural Resources Conservation Service (NRCS), "Sage Grouse Initiative" at http://www.sagegrouseinitiative.com/about/ (last accessed June 10, 2019). For a more detailed discussion of the Sage Grouse Initiative, see the NRCS publication *Sage Grouse Initiative: Success on the Range* at http://www.sagegrouseinitiative.com/wp-content/uploads/2015/07/SGI-Success-on-the-Range-FINAL-LOW-RES-FULL-SPREADS-WITH-HYPERLINKS-FOR-WEB-073015.pdf (last accessed June 10, 2019).

39. Quoted in "USDA, Montana Enter Public-Private Agreement to Enhance Montana's Sage Grouse Conservation Efforts" at http://www.nrcs.usda.gov/wps/portal/nrcs/detail/national/newsroom/releases/?cid=NRCSEPRD377619 (last accessed June 10, 2019).

40. "U.S. Interior Department Denies Endangered Species Act Protection to Sage Grouse," Associated Press, September 22, 2015, at http://www.nbcnews.com/news/us-news/u-s-interior-department-denies-endangered-species-act-protection-sage-n431701 (last accessed 10 June 2019).

41. Though the precise impact is difficult to ascertain, it is reasonable to assume that the threat of regulatory intervention can encourage private property owners to adopt approved habitat conservation plans (HCPs) to avert regulatory intervention, an option specified by the U.S. Endangered Species Act. In an article published in the *Journal of Environmental Economics and Management*, Christian Langpap and Joe Kerkvliet, both of whom are affiliated with Oregon State University, report that an in-depth study they did of voluntary habitat conservation plans (HCPs) on private land indicated that "on average, HCPs have positive effects on endangered species recovery." Such being the case, they suggest that the U.S. Fish and Wildlife Service (FWS) and the National Marine Fisheries Service (NMFS) "should further encourage private parties to develop HCPs, particularly for listed species that currently do not have a plan." They conclude that "HCPs can have a positive impact on species recovery and that, overall, endangered species' chances of survival are greater with an HCP than without" ("Endangered Species Conservation on Private Land: Assessing the Effectiveness of Habitat Conservation Plans," *Journal of Environmental Economics and Management*, July 2012, 64:14). See also Christian Langpap and JunJie Wu, "Voluntary Conservation of Endangered Species: When Does No Regulatory Assurance Mean No Conservation?" *Journal of Environmental Economics and Management*, June 2003, 47:435–457, and Christian Langpap, "Conservation of Endangered Species: Can Incentives Work for Private Landowners?" *Ecological Economics*

(2006), 57:558–572. In the latter article, Langpap argues that "conservation policy for private lands could be improved by relying on a combination of incentives, including financial incentives and assurances, rather than exclusively on the threat of regulation" (558).

42. U.S. Fish and Wildlife Service, "Endangered Species Act Timeline" at http://www.fws.gov/endangered/laws-policies/timeline.html (last accessed June 10, 2019).

43. Defenders of Wildlife, "Basic Facts about Bald Eagles" at http://www.defenders.org/bald-eagle/basic-facts (last accessed June 10, 2019).

44. Center for Biological Diversity, "Restoring the Gray Wolf" at http://www.biologicaldiversity.org/campaigns/gray_wolves/ (last accessed September 3, 2016). See also Clyde Haberman, "For Gray Wolves, a Success Story Not Without Detractors," *New York Times,* 2 November 2014, and Defenders of Wildlife, "Gray Wolf Recovery Infographic" at http://www.defenders.org/the-war-on-wolves/infographic (last accessed June 10, 2019).

45. University of Montana, "Grizzly Bear Ecosystems" at http://www.cfc.umt.edu/grizzlybearrecovery/grizzly-bears/ecosystems.php (last accessed June 10, 2019). The grizzly population, however, is sparse in the other three grizzly bear ecosystems—the North Cascades, the Selkirk, and the Cabinet-Yaak ecosystems.

46. U.S. Fish and Wildlife Service, "Endangered Species Act Timeline."

47. David S. Wilcove, Michael J. Bean, Robert Bonnie, and Margaret McMillan, *Rebuilding the Ark: Toward a More Effective Endangered Species Act for Private Land*, December 5, 1996, at http://omnilearn.net/esacourse/pdfs/Rebuilding_the_Ark.pdf (last accessed June 10, 2019).

48. Haberman, "For Gray Wolves, a Success Story Not Without Detractors."

49. See, for example, Bill Paul, Ed Bangs, Carter Niemeyer, and Elizabeth Harper, "Wolf Depredation," January 2014 at http://www.defenders.org/press-release/defenders-shifts-focus-wolf-coexistence-partnerships (last accessed June 10, 2019).

50. "Defenders Shifts Focus to Wolf Coexistence Partnerships," August 20, 2010, at http://www.defenders.org/press-release/defenders-shifts-focus-wolf-coexistence-partnerships (last accessed June 10, 2019).

51. Minnesota Department of Agriculture, "Guide for Minnesota Farmers and Ranchers Living in Wolf Territory" at http://www.mda.state.mn.us/grants/disaster/wolf.aspx (last accessed September 3, 2016).

52. "Oregon Department of Agriculture Will Compensate Ranchers for Livestock Killed by Wolves" at http://www.oregonlive.com/environment/index.ssf/2011/12/oregon_begins_compensation_pro.html last accessed June 10, 2019).

53. "Compensation Rules for Depredation Incidents" at http://wdfw.wa.gov/conservation/gray_wolf/livestock/compensation.html (last accessed October 21, 2015).

54. "Montana Livestock Loss Board" at http://liv.mt.gov/llb/default.mcpx (last accessed October 21, 2015).

55. Endangered Species Act of 1973, Section 4 (3)(A).

56. For a definitive list of threatened species worldwide, see the International Union for Conservation of Nature *Red List of Threatened Species* at http://www.iucnredlist.org/ (last accessed June 10, 2019).

57. See, for example, Elizabeth Kolbert, *The Sixth Extinction: An Unnatural History*, (New York: Henry Holt and Company, 2014), 265–269 and Denise Chow, "One Million Species under Threat of Extinction because of Humans, Biodiversity Report Finds," *NBC News*, May 6, 2019 at https://www.nbcnews.com/mach/science/1–million-species-under-threat-extinction-because-humans-report-finds-ncna1002046 (last accessed June 10, 2019). Protecting endangered species in countries that have effective environmental laws and programs, it might be added, is a far easier task that in countries lacking these tools of protection.

Part II

Productivity and Sustainable Agriculture

Chapter Four

"A Whole Arsenal of Agricultural Chemicals"

In Wendell Berry's novel *Jayber Crow*, Athey Keith was one of the best farmers in the neck-of-the-woods in rural Kentucky where the novel is set. He owned five hundred acres, a fairly sizable farm for the first half of the twentieth century. Almost all of it was bottom land near the river, arable except for the hollows and timber patches which occupied seventy-five or eighty acres. Athey rotated the crops on the various fields he farmed—tobacco and corn followed by wheat or barley and then by clover and grass. He had livestock—cattle, sheep, and hogs—and in the long barn behind his house he kept a dozen or so good work mules, which he harnessed up when spring came to pull the farm implements he used to till the soil. Athey took good care of his land, never overgrazing the pasture and, in the fields he planted crops, replenishing the nutrients in the soil by working in the manure his cattle, sheep, hogs, and mules generously contributed to the ecosystem. He believed in maintaining reserve capacity by not overtaxing the land. "Wherever I look," he once observed, "I want to see more than I need, and have more than I use."[1]

Athey and his wife, Della, had but one child—a daughter they named Mattie, who in time was to become Mattie Chatham. Athey and Della did not particularly take a shine to Troy Chatham, who was a show-off and a braggart. At first, they refused to give their consent to the marriage but in time relented because they realized that once their daughter turned twenty-one, she would marry Troy with or without their blessing. Troy and Mattie were married in late autumn of 1945, shortly after Mattie's twenty-first birthday. It was not a match made in heaven, for Troy was cut from entirely different cloth—entirely different cloth not just from Mattie but from Athey and Della as well.[2]

Troy did not share Athey's love for the land. He viewed land merely as something to be exploited. "Never let a quarter's worth of equity stand idle," he asserted, "Use it or borrow against it."[3] Viewing himself as part of the wave of the future, which in a certain sense he was, Troy believed that bigger is better—more land on which to raise more corn and other crops in the hope of making more money. Scarcely a year after he and Mattie were married, he bought a tractor and several tractor-drawn farm implements, using most of what he made from that year's crop as a down payment and borrowing the rest from a bank in town. "We need to grow more corn," he announced to Athey. "If you raise more corn," Athey responded, "you'll have to buy fertilizer," knowing that Troy would need to borrow against the next year's crop to pay for the fertilizer, which he proceeded to do. While Athey did agree to let Troy plant ten more acres of corn, Troy wanted more—much more. He rented land from neighbors so that he could raise still more corn, working late into the night on his tractor (which, unlike the mules, had headlights and did not need to rest at the end of the day), leaving Mattie to do the barn chores.[4]

When the sands of time began catching up with Athey and Della, they moved to Port William, the small town nearby, where they had a garden large enough to provide the vegetables they needed, a milk cow, and some chickens for eggs. Athey kept a couple of the older mules with which he could not bring himself to part company, using them to plow and cultivate the garden when it needed to be plowed or cultivated. Athey was the first to pass away, followed ten years later by Della. With Athey and Della gone, Troy pretty much did what he wanted with the once fertile farm. Though it was technically owned by Mattie, who had inherited the land, she was powerless to stop him. Troy's mantra was "modernize, mechanize, specialize, grow." He tore down the fences, plowed up everything that could be plowed, and, abandoning the careful crop rotation that had been Athey's way of farming, piled on artificial fertilizers in an effort to boost production, which worked for a while. But in time he ruined the land and ruined himself financially, leaving Mattie and their children with very little in the way of support, financial or otherwise.[5]

The story of Athey and Della, Mattie and Troy, of course, is fictitious. But like many works of fiction, *Jayber Crow* is rich in social commentary (notwithstanding Berry's claim to the contrary in the front piece). Agriculture in the United States *was* substantially transformed in the twentieth century, much along the lines that Berry uses fiction to recount in *Jayber Crow*. Horses and mules gave way to tractors. And in turn, tractors such as the iconic gasoline-powered two-cylinder John Deere Model A, which could pull a three-bottom plow on level ground, gave way to huge diesel-powered tractors which can pull ten-bottom plows capable of covering nearly twenty feet a pass. In time, farming changed from being a way of life to being a business, with artificial fertilizers, herbicides, and other pesticides used in substantial amounts in an effort to boost crop yields, and, by so doing, profits, though, as Troy Chatham discovered in *Jayber Crow*, the latter has not always been successful.

HUGE INCREASES IN YIELDS PER ACRE

The increases in yields per acre in last half of the twentieth century and the first part of the current century are phenomenal. Take, for example, corn. At the conclusion of the Civil War, corn yield in the United States averaged 30 bushels per acre. With some yearly variations related to weather conditions, it remained at that level for the better part of a century.[6] Yields began increasing with the introduction of hybrid seed corn in 1933. The major increases, however, did not occur until after World War II when methods of farming changed substantially. John Fraser Hart, who taught at the University of Minnesota for a number of years, observes:

> Hybrid corn enabled [farmers] to double or even triple their yields, but also forced them to start using a whole arsenal of agricultural chemicals. Before World War II they had been able to maintain acceptable levels of productivity with no more than barnyard manure . . . . Between 1949 and 1969 Corn Belt farmers increased their . . . use of fertilizer sixfold to take full advantage of the potential productivity of hybrid corn. Batteries of gleaming white tanks of anhydrous ammonia—a lethal gas that is a superb source of nitrogen if it is injected into the soil properly—have become a common sight in dealers' yards at the edge of almost every small town in the Corn Belt.[7]

Hart notes, "In 1950 farmers in the Corn Belt were wondering if they would ever live to see yields of 100 bushels an acre, but by 1985 they were feeling sorry for themselves if they failed to get at least 125 [bushels an acre]."[8] By the end of the first decade of the twenty-first century, the *average* corn yield was more than 150 bushels per acre[9] with yields in excess of 200 bushels per acre being achieved by some farmers when the rain came when needed and growing conditions were favorable.

Prior to this transformation of agriculture, most farms were diversified operations with farmers raising a variety of crops on a rotating basis, along with the amount and types of livestock the farm could sustain, most typically cattle, pigs, and chickens. The manure from the livestock (a manure spreader was one of the most important farm implements, though as dealers were fond of saying, it was the one piece of equipment they did not stand behind) and rotation of crops maintained the fertility of the soil.

In the traditional way of farming, weeds were controlled by cultivating. The advent of herbicides that could knock out noxious weeds eliminated the need for frequent cultivation. This not only saved time but enabled reducing the space between rows. It also enabled planting individual seeds a few inches apart in the row, rather than in hills, which further contributed to productivity. The result was more stalks of corn per acre, which translated into more bushels of corn per acre.

While the increases in yield per acre for wheat and other crops are not as dramatic as those for corn, they are still quite significant. In 1949, U.S. wheat yields were essentially the same as those seventy-five years earlier—roughly 15 bushels per acre. But then with increased use of artificial fertilizers and herbicides, yields nearly doubled from 15 bushels per acre to almost 30 bushels per acre in 1976.[10] By 2010, yields averaged 46 bushels per acre.[11] Yields of other crops such as soybeans, oats, and barley similarly increased in the post-World War II years, a trend that continues in the present century. The common thread running through all of this is the use of artificial fertilizers, herbicides, and other pesticides.[12]

SOME BENEFITS OF INCREASED PRODUCTIVITY

This transformation of agricultural production does have some benefits. Food is abundant in the United States and, notwithstanding what grocery shoppers often believe to be the case, is relatively inexpensive. Americans spend approximately 6 percent of their household income on food (less than 10 percent when eating out is included), significantly less than any other country in the world.[13] Current U.S. expenditures on food compare favorably to the 29.7 percent of household income Americans spent on food in 1950 and the 42.5 percent spent on food in 1901.[14] (Farmers receive less than 20 percent of what Americans spend on food today; processing, distribution, advertising, and other costs, along with profits that accrue to processors, distributors, and retailers, account for the rest.[15])

There are other benefits as well resulting from high levels of productivity. U.S. exports of corn, wheat, soybeans, and other agricultural commodities, which grew dramatically in the last half of the twentieth century,[16] help offset imports of television sets, clothing, and other goods, thus lessening the U.S. balance of payments deficit.[17] And the abundance of corn has contributed to the growth of the ethanol industry, reducing U.S. dependence on imported oil. By 2010, nearly 40 percent of the U.S. corn crop was used to produce ethanol.[18]

THE CHANGING FABRIC OF RURAL AMERICA

All of this has come with significant costs, however, among them costs to the social fabric of rural America. In *Jayber Crow*, Port William gradually withers away. When Will Johnson died, his blacksmith shop closed and never reopened. When Milton Burgess died, which was about the same time that Athey Keith passed away, the general store closed. Efforts to find someone to buy it were unsuccessful. Three years later, school officials closed the local school and arranged to have the twenty or so children who remained bussed to a school in another town. The following year, the old Port William Hotel, which had become an apartment house for the elderly, burned to the ground. When the interstate highway was built, those remaining in Port William found it convenient to drive to Louisville to do their shopping.[19]

The story of Port William is the story of much of rural America in the last half of the twentieth century and the first part of the current century. Locally-owned grocery stores, hardware stores, and apparel stores have closed as the remaining residents of the small towns that still exist get in their cars and drive a half hour or so to the nearest Wal-Mart store, where they have a greater choice of merchandise at lower prices, or to the nearest supermarket, which often sells groceries retail for less than small locally-owned grocery stores pay wholesale for items to stock their shelves. Such are the economies of scale.

The demise of small-town rural America is not just the story of automobiles and increased mobility. As farms have gotten larger, there are fewer people involved in agriculture and the supportive services related to agriculture. At the dawn of the twentieth century, 41 percent of the U.S. labor force worked on farms. A century later, that had declined to 1.9 percent of the labor force. Since 1900, the number of farms in the United States has declined by 63 percent while the average farm size has increased 67 percent.[20]

The smaller farms that remain, it should be added, are seldom self-sustaining. A one-liner making the rounds in many rural communities is that a successful farmer is someone whose wife (or husband) works for the county.

Powerful economic forces have driven this change. Adjusted for inflation, market prices for wheat, corn, and soybeans are often lower than they were in 1920.[21] Meanwhile, prices for farm machinery, fertilizer, and virtually everything else that farmers need have gone up, resulting in paper-thin profit margins, or in some years, no profit at all. Writing in the latter part of the twentieth century, Hart observed:

> One of the few things that has hardly changed in the Corn Belt since World War II has been the price the farmer has received for his crops. The price of corn in real dollars, for example, has remained as low as it has been at any time since the Civil War, except for a brief but spectacular upsurge in the early 1970s. The cost of growing corn has continued to rise, and the farmer's profit on each bushel of corn, or on each unit of anything else he produces, has been steadily reduced as the gap has narrowed between rising costs and stable or falling prices.

> He has had to compensate for his low profit per unit by producing more units, which is another way of saying that he has had to increase his volume of production to remain in business. The new technology has enabled him to produce more units from each acre of land, but he has also had to increase his acreage. The catchword of the Corn Belt since World War II has been "Get Bigger or Go Under."[22]

As farms grew in size, there came to be increased specialization. The old family farm that produced several different crops and had chickens, pigs, cattle, and other livestock is by and large a thing of the past. Many farmers today have no livestock at all but instead raise only corn, soybeans, or other crops to sell—often just one or two crops. Livestock production has shifted from small diversified farms to large animal factory farms, many owned by investors who are not themselves directly involved in agriculture. These include large dairy, hog, and chicken confinement operations, where the animals, instead of being allowed to wander about in open fields or large pens, are kept indoors where they are fed and watered in confined spaces, with, in many cases, their feed laced with antibiotics and other chemicals designed to prevent disease and promote growth. A 2008 report of the Pew Commission on Industrial Farm Production, a joint project of the Pew Charitable Trusts and Johns Hopkins Bloomberg School of Public Health, notes, "Intensive animal production and processing have brought about significant change in American agriculture over the last two decades. The current trend in animal agriculture is to grow more in less space, use cost-efficient feed, and replace labor with technology to the extent possible."[23]

The result? More meat and eggs at lower prices. But this comes at a cost. Animal rights activists are justifiably concerned about the way that animals are treated in these confinement operations. For example, chickens in egg factories are crammed in small cages with little room to move, the tips of their beaks often having been removed to reduce injury from aggressive behavior resulting from overcrowding. The Pew Report observes, "While industrial farm animal production has benefits, it brings with it growing concerns for public health, the environment, animal welfare, and impacts on rural communities."[24]

Because the old diversified family farm has assumed iconic dimensions in American folklore, it is easy to become nostalgic about what has been lost. The same is true for the small towns that once dotted rural America, some of which no longer exist while others exist only in diminished form. There is no question that a way of life—and with it a certain quality of life—has, with rare exception, been irretrievably lost.[25] Most sons and daughters of farmers (myself included) have long since left the farms on which they were raised for jobs in cities or for other lines of work, both as a matter of economic necessity and as a matter of choice. Indeed, one of the reasons that farms are getting larger today is that as farmers reach retirement age, they often sell their land to larger farming operations because their sons and daughters have no desire to return to the farm.

ENVIRONMENTAL COSTS

The impact on the social fabric of rural communities is only one of the costs of the technological transformation of agriculture. Equally significant—arguably even more significant—are the environmental costs. These include topsoil erosion, ground water contamination, and contamination of streams, rivers, and lakes. When farmers moved away from a pattern of rotating crops and instead planted corn, soybeans, and other crops from fence line to fence line, there were huge environmental costs as tons of topsoil blew or washed away because it no longer had cover to protect it. When farmers began using large quantities of artificial fertilizers, herbicides, and other pesticides, these chemicals ended up in ground water, including the aquifers from which wells drew drinking water, and in the runoff going to streams, rivers, and lakes. Extensive use of antibiotics in the feed of meat-producing animals left antibiotic residues in the meat consumed by humans, contributing to the growth of antibiotic resistant strains of salmonella and other bacteria. Each of these areas of concern merits more detailed discussion.

TOPSOIL EROSION

In recent years, there has been significant progress in reducing topsoil erosion, but much remains to be done with topsoil erosion in many areas continuing to be a very serious problem. The *2015 National Resources Inventory* conducted by the National Resources Conservation Service (NRCS), a branch of the U.S. Department of Agriculture (USDA), reports that soil erosion on cultivated non-federal cropland in the United States decreased 34 percent between 1982 and 2015, with both water and wind erosion showing significant decreases.[26] Measures that have been taken to reduce topsoil erosion will be discussed in greater detail in the next chapter.

GROUND WATER AND SURFACE WATER CONTAMINATION

Contamination of ground water (water below the surface of the land) and surface water (streams, rivers, ponds, and lakes) is another environmental cost of the productivity revolution in agriculture. Extensive use of artificial fertilizers containing nitrogen, atoms of which join to oxygen atoms to form nitrates, can result in increased nitrate levels in both ground and surface water.[27] (The three numbers that appear on bags of fertilizer in the United

States indicate the percentage of nitrogen, phosphorus, and potassium, respectively; e.g., a bag of fertilizer labeled 5–10–5 must contain a minimum of 5 percent nitrogen, 10 percent phosphorus, and 5 percent potassium.[28]) Large concentrations of animal waste that are not adequately contained can also lead to increased nitrate levels and other contaminants in ground or surface water.[29] (Chemically, plant nutrients from artificial fertilizers and animal wastes are identical.[30])

Nitrates, which are water soluble, are naturally present in soil, water, and food and are a normal component of the human diet. When ingested, nitrates, which have three oxygen molecules, convert to nitrites by losing one of their oxygen molecules. (Nitrates and nitrites are also used as food preservatives; both are widely used to maintain the color and extend the shelf life of processed meats.) Nitrites react with hemoglobin in the blood, causing methemoglobin, which decreases the capacity of the blood to transport oxygen. Healthy individuals have low levels of methemoglobin in their blood (0.5 to 2.0 percent). When this level increases to 10 percent, the skin and lips can take on a bluish tinge (cyanosis), a symptom of methemoglobinemia. High levels of methemoglobin can be lethal. Because infants are much more sensitive to nitrates and nitrites than are adults (the stomach acid of infants is not as strong as in older children and adults, which results in an increase in the bacteria that convert nitrate to nitrite[31]), virtually all deaths from nitrate/nitrite poisoning (methemoglobinemia) have involved infants.[32]

The U.S. Environmental Protection Agency has set the maximum contaminant level goal (MCLG) for nitrates in drinking water at 10 parts per million (ppm).[33] In the United States, nitrates are present in most surface water and ground water at levels below 4 ppm, though levels exceeding 20 ppm have been found to be present in approximately 3 percent of surface water and 6 percent of ground water.[34] (While nitrate pollution from agriculture has received considerable attention and is of particular concern for this discussion, excessive or otherwise inappropriate use of artificial fertilizers on lawns in suburban and urban areas also contribute to pollution of ground and surface water.)

While much of the concern about water contamination resulting from modern agricultural practices has focused on nitrates, there are also other problem areas. Among these is phosphorus contamination. Phosphorus, which is not as soluble as nitrate, attaches itself to soil particles and enters streams, rivers, and lakes when these soil particles are washed away via soil erosion, where, along with nitrates, they contribute to eutrophication—the excessive growth of algae which, when they die and decompose, deplete the oxygen supply to the detriment of fish and other species.[35]

Pesticides and herbicides used in agricultural operations are other contaminants in ground and surface water. A study conducted as part of the National Water Quality Assessment (NAWQA) Program of the United States

Geological Survey (USGS) over a ten-year period (1992–2001) discovered one or more pesticides in 97 percent of the streams surveyed and 61 percent of the shallow ground water. In the vast majority of cases, the concentrations were below the level that posed a threat to human health. Yet 9.6 percent of the streams surveyed and 1.2 percent of the shallow ground water tested exceeded these levels. One finding of particular concern was that measurable levels of DDT could still be detected in sediment in stream beds in some parts of the country, notwithstanding the fact that the use of DDT has been banned in the United States since 1972.[36]

A North Dakota study determined that factors related to potential contamination of ground water by agricultural pesticides include depth to the aquifer (an underground layer of water-bearing permeable rock, gravel, sand, or silt), water flow direction (including in the aquifer), permeability of soil (sandy soils are far more permeable than clay and hence allow more water to soak through them), organic matter content (soils higher in organic content have a higher filtration potential), and the chemistry of the particular pesticides being used (pesticides that disintegrate rapidly are less problematic than those such as DDT that disintegrate slowly).[37]

Atrazine, an herbicide used to control grass and broadleaf weeds sold by Syngenta AG, based in Basel, Switzerland, has been a matter of controversy in the United States and elsewhere. Atrazine is not banned in the United States. Since 1992 the U.S. Environmental Protection Agency has specified the maximum atrazine level that can be in drinking water with the water still considered safe. More than 1,000 communities, mostly in the Midwest, have spent millions of dollars for filtration systems to reduce the atrazine level in their drinking water.

RISKS TO HUMAN HEALTH RESULTING FROM USE OF ANTIBIOTICS IN LIVESTOCK PRODUCTION

As noted above, feed laced with antibiotics and other chemicals is widely used on animal factory farms to prevent disease and spur growth. Most members of the public are not aware of how extensive the use of antibiotics in livestock production is. A substantial portion of the antibiotics sold in the United States—many of these the same drugs that are used to fight infections in humans—are used by livestock producers. Some of these antibiotics are used to treat animals with bacterial infections. However, many of these antibiotics are used for nontherapeutic purposes—that is, given to healthy animals in an effort to prevent them from getting infectious diseases.

Though the data related to the use of antibiotics in food animals is somewhat sketchy, what is clear is that there is an alarming increase in the incidence of antibiotic resistant bacteria that pose significant health risks for humans. The World Health Organization (WHO) reports that there are 440,000 new cases of multidrug-resistant tuberculosis each year, causing at least 150,000 deaths. WHO report further notes that a high percentage of hospital-acquired infections are caused by highly-resistant bacteria such as methicillin-resistant *Staphylococcus aureus (MRSA)*.[38]

The Centers for Disease Control and Prevention (CDC) reports that in the United States, "Antibiotic resistance is one of the biggest public health challenges of our time. Each year in the United States, at least 2 million people get an antibiotic-resistant infection and at least 23,000 people die each year as a result of these infections."[39]

This alarming situation is not solely the result of the use of antibiotics in livestock production. There are other causes as well, among them misuse of antibiotics to treat diseases in humans such as viral infections for which antibiotics are not medically indicated, and failure of patients to take all of the antibiotic pills prescribed. (It is not unusual for patients to discontinue taking their medication when they start to feel better, thereby allowing bacteria that have undergone mutations to live to attack another day.)

Yet, though the data are incomplete, there is reason to believe that agricultural use of antibiotics is a contributing factor. Take, for example, Russ Kremer, a Missouri hog farmer who developed a blood infection after being gored by one of his pigs. His infection was resistant to six of the seven antibiotics used to treat it.[40] Kremer's doctor told him that he had the same antibiotic resistance as did his pigs.[41]

In a statement that appeared in the *Federal Register* in 2012, the FDA observed, "In regard to antimicrobial drug use in animals, the Agency considers the most significant risk to the public health associated with antimicrobial resistance to be human exposure to food containing antimicrobial-resistant bacteria resulting from the exposure of food-producing animals to antimicrobials . . . ."[42] The Pew Commission report on industrial farm animal production (IFAP) states, "Whatever the direct evidence, it is certain that the exposure of bacteria to antimicrobial agents selects resistant bacteria that can replicate and persist. Such bacteria from IFAP facilities can reach humans through many routes, both direct (through food, water, air, or contact) and indirect (via transmission of resistance in the environmental pool of bacteria)."[43]

In 2013, the Food and Drug Administration (FDA) announced plans to phase out nontherapeutic use of antibiotics in food animals by asking drug companies to voluntarily take measures to ensure that the drugs they sell would only be used for therapeutic purposes supervised by veterinarians.[44] As will be noted in greater detail in the next chapter, additional measures have more recently been taken to address the problem.

NOTES

1. Wendell Berry, *Jayber Crow* (New York: Counterpoint, 2000), 178–179.
2. Ibid., 176–177.
3. Ibid., 179.
4. Ibid., 185.
5. Ibid., 278–279, 334–343.
6. U.S. Department of Energy Alternate Fuels Data Center, "U.S. Corn Productivity and Land Use" at http://www.afdc.energy.gov/data (last accessed June 11, 2019).
7. John Fraser Hart, "Change in the Corn Belt" at http://www.geo.hunter.cuny.edu/courses/geog347/articles/change_belt.pdf (last accessed June 11, 2019).
8. Ibid.
9. Iowa State University Extension, "Corn and Soybeans: Harvested Acreage and Yield per Acre" at http://www2.econ.iastate.edu/outreach/agriculture/periodicals/chartbook/Chartbook2/Tables/Table10.pdf (last accessed June 11, 2019).
10. Jack J. Bond and Dwaine E. Umberger, *Technical and Economic Causes of Productivity Changes in U.S. Wheat Production, 1949–1976*, U.S. Department of Agriculture Technical Bulletin 1598 (1979), 6–7.
11. U.S. Census Bureau, *Statistical Abstract of the United States: 2012*, Table 861 (551). The statistics are for all varieties of wheat—winter wheat, spring wheat, and other varieties.
12. There are also some other factors that account for increased yields, among them improved varieties of the crops being raised and increased use of irrigation.
13. Alyssa Battistoni, "America Spends Less on Food Than Any Other Country," *Mother Jones*, February 1, 2012 at http://www.motherjones.com/blue-marble/2012/01/america-food-spending-less (last accessed June 11, 2019) and Annette Clauson, "Despite Higher Food Prices, Percent of U.S. Income Spent on Food Remains Constant," USDA Economic Research Service (September 2008) at http://webarchives.cdlib.org/wayback.public/UERS_ag_1/20110913223320/http://www.ers.usda.gov/AmberWaves/September08/Findings/PercentofIncome.htm (last accessed November 5, 2012).
14. *100 Years of U.S. Consumer Spending: Data for the Nation, New York City, and Boston*, U.S. Bureau of Labor Statistics (May 2006), 3, 22.
15. "A New Look at Where Our Food Dollars Go," USDA Economic Research Service (June 2011) at http://www.ers.usda.gov/amber-waves/2011–june/data-feature-a-new-look-at-where.aspx (last accessed November 5, 2012).
16. Carolyn Dimitri, Anne Effland, and Neilson Conklin, "The 20th Century Transformation of U.S. Agriculture and Farm Policy," USDA Economic Research Service (June 2005) at http://ageconsearch.umn.edu/bitstream/59390/2/eib3.pdf (last accessed June 11, 2019).
17. "Foreign Agricultural Trade of the United States (FATUS)," USDA Economic Research Service (September 30, 2011) at http://www.ers.usda.gov/data/fatus (last accessed November 5, 2012).
18. Lihong McPhail and Xiaodong Du, "Ethanol Strengthens the Link Between Agriculture and Energy Markets," USDA Economic Research Service (June 2012) at http://www.ers.usda.gov/amber-waves/2012–june/ethanol-strengthens-the-link.aspx (last accessed November 6, 2012).
19. Berry, *Jayber Crow*, 258, 274, 279–282.
20. Dimitri, Effland, and Conklin, "20th Century Transformation of U.S. Agriculture and Farm Policy," 7.
21. Stuart Staniford, "Early Warning: Long Term Crop Prices" (March 24, 2011) at http://earlywarn.blogspot.com/2011/03/long-term-crop-prices.html (last accessed June 11, 2019).
22. Hart, "Change in the Corn Belt," 55.
23. Pew Commission on Industrial Farm Animal Production, *Putting Meat on the Table: Industrial Farm Animal Production in America* (April 29, 2008), 5.
24. Ibid., 7.
25. Some are attempting to recover some portion of this way of life by rejecting more affluent lifestyles in favor of living very simply with a bare minimum of material goods. See,

for example, Ralph Blumenthal and Rachel Mosteller, "Voluntary Simplicity Movement Re-Emerges, *New York Times*, May 18, 2008.
26. U.S. Department of Agriculture, *Summary Report: 2015 National Resources Inventory*, USDA Natural Resources Conservation Service, Washington, D.C., and Center for Survey Statistics and Methodology, Iowa State University, Ames, Iowa (September 2018), 2–8 at https://www.nrcs.usda.gov/Internet/FSE_DOCUMENTS/nrcseprd1422028.pdf (last accessed June 11, 2019).
27. "Human Health Fact Sheet: Nitrate and Nitrite," Argonne National Laboratory, EVS (August 2005). While making fertilizer is the primary use of nitrates, they are also used to make glass and certain types of explosives such as those used by Timothy McVeigh to attack the Alfred P. Murrah Federal Building in Oklahoma City on April 19, 1995.
28. "What Do the Numbers of Fertilizer Packages Mean?" Oregon State University Extension Service at http://extension.oregonstate.edu/gardening/node/679 (last accessed June 11, 2019).
29. "Human Health Fact Sheet: Nitrate and Nitrite."
30. Daryl Bucholz, "Agricultural Sources of Contaminants in Ground water," University of Missouri Extension Service at http://extension.missouri.edu/p/WQ21 (last accessed November 9, 2012).
31. J. R. Self and R. M. Waskom, "Nitrates in Drinking Water," Colorado State University Extension Fact Sheet No. 0.517 (October 2008).
32. "Human Health Fact Sheet: Nitrate and Nitrite."
33. "Basic Information about Nitrate in Drinking Water," United States Environmental Protection Agency at http://water.epa.gov/drink/contaminants/basicinformation/nitrate.cfm (last accessed June 11, 2019). The 10 ppm guideline is for nitrate-N; the equivalent standard for nitrate-NO3 is 45 ppm.
34. "Nitrate and Nitrite in Drinking Water: Background document for development of WHO Guidelines for Drinking-Water Quality" (WHO/SDE/WSH/07.01/16/Rev/1), 3.
35. National Oceanic and Atmospheric Administration (NOAA), "What Is Eutrophication?" at https://oceanservice.noaa.gov/facts/eutrophication.html (last accessed June 11, 2019).
36. "Pesticides in the Nation's Streams and Ground Water, 1992–2001—A Summary," U.S. Geological Survey (March 2006) at http://pubs.usgs.gov/fs/2006/3028/pdf/fs2006–3028.pdf (last accessed June 11, 2019).
37. John Nowatzki, "GIS Application of Potential Contamination of Ground water from Agricultural Pesticides" at http://proceedings.esri.com/library/userconf/proc00/professional/papers/pap198/p198.htm (last accessed June 11, 2019). GIS refers to the Geographic Information System, a software package useful for assessing the potential for ground water contamination.
38. World Health Organization, "Antimicrobial Resistance" at http://www.who.int/mediacentre/factsheets/fs194/en/# (last accessed June 11, 2019).
39. Centers for Disease Control and Prevention (CDC), "Antibiotic/Antimicrobial Resistance (AR/AMR)" at https://www.cdc.gov/drugresistance/ (last accessed June 11, 2019).
40. Karen McVeigh, "Scientists: Overuse of Antibiotics in Animal Agriculture Endangers Humans," *Guardian*, September 19, 2012.
41. McVeigh, "Overuse of Antibiotics in Animal Agriculture."
42. Food and Drug Administration, "New Animal Drugs; Cephalosporin Drugs; Extralabel Animal Drug Use; Order of Prohibition," *Federal Register* (January 6, 2012) 77, 4: 738.
43. Pew Commission, *Industrial Farm Animal Production in America*, 16.
44. Catharine Paddock, "FDA to Phase out Non-Therapeutic Use of Antibiotics in Food Animals," *Medical News Today*, December 12, 2013, at http://www.medicalnewstoday.com/articles/270010.php (last accessed June 11, 2019).

Chapter Five

New Ways to Farm Profitably While Protecting the Environment

A fairly typical definition of sustainable agriculture is that of the Food and Agriculture Organization (FAO) of the United Nations, which defines sustainable agriculture as "the management and conservation of the natural resource base, and the orientation of technological and institutional change in such a manner as to ensure the attainment and continued satisfaction of human needs for present and future generations." The FAO definition adds, "Such development conserves land, water, plant and animal genetic resources, is environmentally non-degrading, technically appropriate, economically viable and socially acceptable."[1]

The National Institute of Food and Agriculture, which is part of the U.S. Department of Agriculture, notes that in the United States, the legal definition of sustainable agriculture, as specified by U.S. Code Title 7, Section 3103, states:

> [Sustainable agriculture] means an integrated system of plant and animal production practices having a site-specific application that will over the long term:
>
> - Satisfy human food and fiber needs.
> - Enhance environmental quality and the natural resource base upon which the agriculture economy depends.
> - Make the most efficient use of nonrenewable resources and on-farm resources and integrate, where appropriate, natural biological cycles and controls.
> - Sustain the economic viability of farm operations.
> - Enhance the quality of life for farmers and society as a whole.[2]

A paper published by the Agricultural Sustainability Institute at the University of California—Davis emphasizes the importance of stewardship:

> Sustainability rests on the principle that we must meet the needs of the present without compromising the ability of future generations to meet their own needs. Therefore, *stewardship of both natural and human resources* is of prime importance. Stewardship of human resources includes consideration of social responsibilities such as working and living conditions of laborers, the needs of rural communities, and consumer health and safety both in the present and in the future. Stewardship of land and natural resources involves maintaining or enhancing this vital resource base for the long term.[3]

In an interview a few years ago, Sally Puttmann, who along with her husband operated a diversified farm in Iowa on which they raised corn, soybeans, hogs, and cattle long after most other farmers opted for specialization, used words both simple and eloquent when she characterized sustainable agriculture as agricultural practices "that will not only economically sustain the family that farms the land, but that will enhance the land long after [that] family is gone."[4] This might well be one of the best definitions of sustainable agriculture.

ORGANIC FARMING

For many, organic farming is the gold standard of the sustainable agriculture movement.[5] It involves farming the way that Athey Keith did in *Jayber Crow*—using crop rotation to maintain the fertility of the soil and working in barnyard manure to enrich it, while eschewing the use of herbicides, pesticides, and artificial fertilizers.

Because organic foods usually command a premium when sold, some type of certification system is necessary to ensure that what is sold as organic really is organic. In 1990, the U.S. Congress passed the Organic Foods Production Act (OFPA), which required the U.S. Department of Agriculture (USDA) to develop national standards for organically produced agricultural products.[6] USDA's National Organic Program (NOP) establishes the requirements that organic production and handling operations must meet in order to be certified by USDA-accredited certifying agents (which include USDA-approved state and private organizations).[7] Similar certification procedures have been established in other countries.[8]

According to statistics compiled by the Organic Trade Association, sales of organic products in the United States increased from $23 billion in 2009 to more than $52 billion in 2018, with more than 90 percent of the sales involving organic food products. In 2018, organic food products were 5.9 percent of all food sold in the United States. Organic fruits and vegetables continue to be the largest organic category, followed by dairy products. [9]

Organic farming has much to recommend it. But if that is the case, why is organic farming not universally practiced? The answer to this question has to do with some basic economic realities. Organic farming tends to be more labor intensive than farming operations that use artificial fertilizers, herbicides, and other pesticides. It is also difficult, if not impossible, to match the high yields achieved by farmers using the modern arsenal of agricultural chemicals. These two factors, along with the cost of maintaining the certification process, result in higher costs of production for organic produce, grains, eggs, meat, and other organic products. This means that organic foods need to command higher prices if organic operations are to be economically viable. For example, a supermarket in Moline, Illinois, which offers both organic and non-organic produce, sells non-organic Honeycrisp apples for $3.99 a pound while organic Honeycrisp apples are $4.99 a pound—a 25 percent premium for organic apples. The same store sells eggs produced by non-organic animal factory farms for $1.99 a dozen, eggs produced by free range operations that do not restrict chickens to cages for $3.99 a dozen, and organic free range eggs for $5.49 a dozen.[10] Because of the more labor intensive nature of organic agriculture, economies of scale, if indeed they could be achieved, are not likely to narrow in any significant way this price differential.[11]

Those strongly committed to environmental preservation, all of whom are to be commended for their idealism, might be inclined to say that everyone should be willing to pay the extra cost for organic produce, grains, eggs, and meat. This is a difficult argument to make, though, to low-income families struggling to survive, to middle-class families trying to scrape up enough money to send their daughters and sons to college, to recent college graduates who are carrying heavy debt loads, and to others who are financially stressed. The reality is that the market for organic produce, grains, eggs, and meat is likely to remain a niche market serving those who are more affluent and have the discretionary income to pay the organic premium (and are willing to do so). This in no way suggests that organic farming is insignificant. For farmers who successfully serve the organic market, it is very significant, as well as for those who purchase organic products. However, even as organic sales increase, realism suggests that organic produce, grains, eggs, meat, and other products are likely to remain a relatively small portion of the total market.

OTHER APPROACHES TO SUSTAINABLE AGRICULTURE

The economic realities that limit the potential of organic farming do not mean that sustainable agriculture is not possible. Scientists at the Leopold Center for Sustainable Agriculture at Iowa State University and elsewhere, including agricultural colleges in other countries, have been instrumental in helping develop agricultural techniques that minimize the need to use artificial fertilizers, herbicides, and other pesticides—techniques which many farmers have adopted. There are other positive developments as well. Environmentally-conscious farmers are using agricultural chemicals that decompose more rapidly, thereby reducing the risk of ground and surface water contamination. Erosion control techniques that help prevent water and wind erosion include allowing grass to grow in gullies where streams of water gush down during heavy rains or when snow is melting, planting buffer zones of grass and bushes along streams and draining ditches to filter out soil carried by runoff, and adopting no-till and low-till methods cultivation that leave crop residue on the surface of the soil. Each of these developments merits more detailed comments.

INTEGRATED PEST MANAGEMENT (IPM)

Farmers used to apply substantial amounts of pesticides and herbicides to make certain that crop-threatening insects and weeds did not have a chance of surviving. The result was overuse of herbicides and other pesticides, damaging both the environment and farmers' pocketbooks. Most farmers today are far more sophisticated about these matters, using knowledge gleaned from scientific research and techniques developed by scientists at land-grant universities and elsewhere to apply herbicides and other pesticides when they will be the most effective, thereby reducing the amount needed to get the job done. The U.S. Environmental Protection Agency maps out a four-tiered approach to IPM:

- *Identifying Action Thresholds.* "Sighting a single pest does not always mean control is needed. The level at which pests will become an economic threat is critical to guide future pest control decisions."
- *Monitoring and Identifying Pests.* "IPM programs work to monitor for pests and identify them accurately, so that appropriate control decisions can be made . . . ." The statement adds, "This monitoring and identification removes the possibility that pesticides will be used when they are not really needed or that the wrong types of pesticide will be used."
- *Prevention.* The EPA statement suggests that the first line of defense against pests is taking preventive measures to prevent them from becom-

ing threats. "In an agricultural crop, this may mean using cultural methods, such as rotating between different crops, selecting pest-resistant varieties, and planting pest-free rootstock. These control methods can be very effective and cost efficient and present little or no risk to people or the environment."

- *Control.* "Once monitoring, identification, and action thresholds indicate that pest control is required, and preventive methods are no longer effective or available, IPM programs then evaluate the proper control method both for effectiveness and risk." The EPA statement continues, "Effective, less risky pest controls are chosen first, including highly targeted chemicals, such as pheromones to disrupt pest mating . . . ." If further action is necessary, "additional pest control methods would be employed, such as targeted spraying of pesticides. Broadcast spraying of non-specific pesticides is a last resort."[12]

In short, IPM is smart farming. Farmers today use their heads far more than their backs. In the United States and other developed countries, computers, and Global Positioning System (GPS) have in substantial measure replaced pitchforks and grain scoops as tools of the trade. Take, for example, the use of computers to calculate when pesticides should be applied. When data such as the number of rainy days and temperatures for these days are entered, sophisticated computer programs can indicate what the optimal time will be for applying various pesticides.[13] This enables farmers to apply pesticides when they will do the most good, getting the job done with a lesser amount than would be the case if the farmer did not take into consideration factors such as the temperature relevant to pests' life cycles—good news both for the environment and for farmers' pocketbooks (and, of course, bad news for pests).

GPS AND PRECISION AGRICULTURE

To the outside observer, it might seem strange that many farmers are installing GPS receivers on their tractors. Don't farmers driving their tractors around in their fields know where they are? Of course they know where they are. Getting lost is not something savvy farmers need to worry about. But that is not why they install GPS receivers on their tractors.

GPS was developed by the U.S. Department of Defense to provide precise targeting information for missiles and other weapons system. The system involves 24 satellites (with three backups) placed in six equally-spaced orbital planes. The system is operated and maintained by the U.S. Air Force. The satellites emit radio signals in which the time and location of the transmitting satellite are embedded. Each satellite circles the earth twice each day.

This ensures that there are at least four GPS satellites in view from virtually any place on the earth.[14] Intercepting signals from three or more GPS satellites, be it by GPS receivers on tractors or receivers built into missile guidance systems, enables using triangulation to determine the precise location of the receiver intercepting the signals.[15]

Precision farming involves using GPS to divide fields into segments, the coordinates of which are recorded. Carefully-labeled soil samples from each segment (typically collected by tubes inserted in the soil that extract a core of soil down to the depth that the roots of the crop that will be planted are likely to reach) are sent to laboratories, where technicians determine nutritional characteristics of the soil—for example, how much nitrogen, phosphorus, and potassium are present in each sample. This data is then entered into computers that determine how much fertilizer is needed in each segment to bring the nutritional values up to the level for optimal productivity. Similar studies can be done to determine the extent to which noxious weeds or crop-threatening insects are present in each segment.[16]

The next step is using computerized variable-rate fertilizer spreaders and pesticide applicators, with GPS determining precise location and the computer varying the application rate based on the data that has been collected for each segment. Thus, for example, if soil testing has determined that a particular segment of a field has a significant nitrogen deficiency, more nitrogen will be applied in that segment while less nitrogen will be applied in segments that are less deficient in nitrogen.[17] This minimizes application of excessive nitrogen which, when it is not used by the crops that are planted, often ends up in surface and ground water. The same applies to herbicides and other pesticides. Combining precision farming with IPM minimizes the use of herbicides and other pesticides, thus reducing the environmental impact of them.

The downside of GPS and other precision agriculture technologies is that they are expensive, though if all goes well, reduced fertilizer and pesticide costs and increased yields offset these costs.

At first glance, this might appear to suggest that the use of precision agriculture technology is limited to large farming operations in view of the fact that it is quite expensive. However, that is not the case. The possibility of hiring farm suppliers to apply fertilizers and pesticides enables smaller farming operations to benefit from the new technologies as well. Remember the "gleaming white tanks of anhydrous ammonia" sitting in "dealers' yards at the edge of almost every small town in the Corn Belt" to which John Fraser Hart made reference? [18] Most of them were not there for purchase by farmers. Rather, farmers typically rented them and often hired a skilled operator affiliated with the dealer to use them to apply the fertilizer. That is precisely what is happening today with respect to GPS-linked computerized variable-rate applicators of fertilizers, herbicides, and other pesticides.

HERBICIDES AND OTHER PESTICIDES THAT DECOMPOSE MORE RAPIDLY THAN OLDER VARIETIES

As noted in the previous chapter, atrazine, an herbicide that has been quite widely used in the United States and other countries, can result in ground and surface water contamination, requiring municipal water systems to install charcoal filtration systems to bring atrazine levels in drinking water down to levels deemed safe for consumption. It is for this reason that most farmers in the United States have significantly reduced their use of atrazine (small amounts are still included in some herbicide mixtures) and instead have shifted to herbicides such as glyphosate that decompose more rapidly, thereby minimizing the environmental impact.[19] Glyphosate does not leach into water systems but instead bonds tightly to soil particles, where it degrades within weeks to form harmless byproducts. Though glyphosate can persist in soil for up to six months, depending on the climate and the type of soil, bacteria in soil usually break it down much sooner.[20] When applied to the leaves of plants, glyphosate kills both broadleaf plants and grasses by preventing the plants from making certain proteins that are necessary for plant growth. It was first registered for use in the United States in 1974.[21] Trade names for glyphosate include Roundup, Rodeo, and Pondmaster.[22]

DDT (dichlorodiphenyltrichloroethane), a potent insecticide that at one time was widely used in the United States and other countries, is now banned in the United States and many other countries but is still used in India and several other countries to control malaria by eliminating *Anopheles* mosquitos, which spread malaria. DDT decomposes slowly and has significant residual environmental costs.[23] *The Stockholm Convention on Persistent Organic Pollutants*, an international treaty that went into effect in 2004, calls upon signatory nations to end agricultural use of DDT but grants an exception for disease control.[24] In 2006, the World Health Organization (WHO), reversing a previous policy, announced that it would actively support the use of DDT to fight malaria."[25] Meanwhile, efforts continue to find other ways of preventing malaria."[26]

Today, farmers in Europe, Canada, Australia, the United States, and other countries have replaced DDT with insecticides which break down more rapidly, among them pyrethroids[27] and dimenthoates,[28] both of which decompose in days or weeks. By carefully monitoring when crop-threatening insects emerge, IPM techniques facilitate using insecticides that have relatively short periods of potency.

NO-TILL AND LOW-TILL CULTIVATION PRACTICES

No-till farming, which disrupts the soil as little as possible by leaving residue from previous crops on the surface when planting the next crop, is gaining popularity throughout the world. It is practiced on farms both large and small, from the Arctic Circle in Finland to countries such as Kenya and Uganda in the tropics, in areas such as western Australia and northern China that have little rainfall and in countries such as Brazil and Chile that have heavy rainfall. No-till techniques are used on sandy soil and on soil with high clay content. South America has the most land dedicated to no-till farming, followed by North America, Australia and New Zealand, Asia, Europe, and Africa.[29] Broken down by country, the United States leads the world in no-till acreage, followed by Brazil, Argentina, Canada, and Australia.[30] In the United States, no-till farming is growing at a rate of about 1.5 percent per year, which is somewhat faster than in other parts of the world. Many farmers selectively use no-till methods on some of their land while using more conventional forms of cultivation on other portions of their land.[31]

In an article entitled "No-Till: the Quiet Revolution" that ran in *Scientific American*, David R. Huggins and John P. Reganold observe, "The age-old practice of turning the soil before planting a crop is a leading cause of farmland degradation. Many farmers are thus looking to make plowing a thing of the past."[32] A newsletter sent out by the Northern Virginia Soil and Water Conservation District in the United States describes no-till agriculture as "good for the soil and the bottom line."[33] Edward Bradley, a no-till farmer and agronomist based in England, notes that no-till farming:

- Reduces soil erosion by leaving crop residues on the surface, rather than plowing them under,
- Improves soil structure by increasing the organic matter in soil,
- Increases the population of earthworms, which help improve the quality of the soil,
- Helps soil retain moisture by increasing the organic matter in the soil,
- Saves fuel by reducing the number of passes that farmers need to make with their tractors to prepare and plant their fields, and
- Improves the oxygen content of soil by not compacting it as a result of multiple passes of heavy equipment.[34]

When using conventional methods of farming, farmers typically started by plowing fields with moldboard plows, which turned the top layer of soil, down to a depth of 8–10 inches, upside down, burying as much as 90 percent of crop residue. This would be followed by using a disk to break up large clods of dirt and tilling the field with a field cultivator to further break up the soil. Prior to planting, farmers would usually run a harrow over the field to smooth it as much as possible. All of this adds up to five passes over the field, culminating in planting the new crop. In contrast, no-till farming involves one pass over the field to apply herbicides, followed by planting.[35] There is an intermediate form of tillage, often called conservation tillage or low-till, that involves working the field with a chisel plow, which leaves approximately 50 percent of the crop residue on the surface, and then cultivating the field with a field cultivator prior to planting.[36]

No-till farming requires the use of special planters and drills that cut narrow grooves through the crop residue on the surface, penetrate the soil to the appropriate depth for planting the seeds for the new crop, place the seeds in appropriate contact with the soil, and close the narrow groove. Fertilizers are often broadcast on the surface, where the nutrients work their way down into the soil. In some cases, crops may need a little more nitrogen the first three to five years of no-till production, though that varies depending on soil and growing conditions. Herbicides are also typically surface applied, ideally at the optimal time for the particular herbicide to be the most effective in killing the weeks being targeted.

ORGANIC FARMING, NO-TILL/LOW-TILL FARMING, AND THE ENVIRONMENT

As noted, many view organic farming as the gold standard of sustainable agriculture. This, however, is not a settled matter. No-till and low-till agricultural practices help prevent erosion by leaving crop residues on the surface, where they protect the soil from wind and water erosion.

While some organic farmers use cover crops to reduce the growth of weeds between the rows of crops that are being raised, others cultivate the soil to control weeds. Some use a combination of both.

Cultivating the soil can make it more susceptible to wind and water erosion. In an article entitled "Is Conservation Farming Better than Organic Farming?" Julia Layton puts it this way: "[Farmers who till the soil] churn up the crop land, pulling up weeds and mixing them into the soil. Many experts attribute the severe erosion problem to this churning up the upper layers of soil before planting. Disturbing the soil cover, loosening it so it's no longer tightly packed, leaves it more susceptible to being displaced by the elements."[37]

No-till and low-till agricultural practices have other environmental benefits as well. By reducing the number of passes that farmers have to make to prepare their fields for planting, they reduce fuel consumption, thereby diminishing the carbon footprint and by so doing, the emission of greenhouse gasses. Moreover, no-till and low-till, by leaving the soil undisturbed, trap more carbon dioxide in the soil, thus reducing the amount that escapes into the atmosphere.[38]

On the other hand, organic farming, by eliminating the use of herbicides and other pesticides, prevents damage to the environment by these and other agricultural chemicals, though, as noted above, improved herbicides and other pesticides that biodegrade more quickly and improved management techniques such as IPM minimize the environmental impact of the use of herbicides and other pesticides in no-till and low-till operations.

So how does this all sort out? In certain climates with certain soil types, organic farming might be the most environmentally friendly while in other areas with other soil types and climactic conditions, no-till might be the best way to go. The bottom line insofar as the environment is concerned is that both organic farming and no-till/low-till farming methods are vastly preferable to conventional tillage. The good news in all of this is that growing numbers of farmers are opting for organic and/or no-till and low-till farming methods.

GENETICALLY MODIFIED ORGANISMS (GMOS) AND SUSTAINABLE AGRICULTURE

The controversy related to genetically-modified foods, which involves claims and counterclaims as to whether they pose a threat to human health, is beyond the scope of this volume. Some types of genetically modified organisms (GMOs), however, have significance for sustainable agriculture, among them herbicide-tolerant (HT) corn and soybeans and genetically-modified crops that are resistant to pests and diseases.

As previously noted, glyphosate, a widely-used herbicide sold under brand names such as Roundup and Rodeo, kills both broadleaf plants and grasses by preventing the plants from making certain proteins that are necessary for plant growth. Genetic engineering techniques have been used, however, to develop varieties of corn, soybeans, and cotton that are tolerant of glyphosate and other herbicides. Planting HT varieties of these crops enables farmers to control weeds by spraying fields of corn, soybeans, and cotton with Roundup and other herbicides without damaging the crop they are raising, which is particularly useful for no-till operations.[39]

By 2014, HT crops accounted for 94 percent of the soybean acreage, 91 percent of the cotton acreage, and 90 percent of the corn acreage in the United States.[40] While HT corn, soybeans, and cotton and other genetically-modified crops are most widely grown in the United States, there is also significant production of genetically-modified crops in Argentina, Brazil, China, and India.[41]

Insect-resistant genetically engineered varieties of crops such as corn and cotton contain a gene from the soil bacterium Bt (*Bacillus thuringiensis*). The insertion of this gene causes the plant to produce a protein that is toxic to specific insects, reducing or eliminating the need to use pesticides. For example, a Bt variety of corn introduced in 2003 is resistant to the corn rootworm, as well as to the European corn borer, to which earlier Bt varieties were not resistant. HT and Bt traits can be "stacked," with the doubly genetically-engineered varieties possessing both herbicide-tolerant and insect-resistant traits. By 2018, Bt varieties accounted for 82 percent of the corn and 85 percent of the cotton planted in the United States.[42]

As noted, the controversy about genetically-modified foods is beyond the scope of this volume. The reason for including an overview of some of the GMOs that are of significance for sustainable agriculture, however, is to remind those involved in the GMO debate that there are environmental benefits resulting from agricultural use of some types of GMOs, benefits that should be included in the overall assessment of GMOs.

REDUCING THE USE OF ANTIBIOTICS AND ARTIFICIAL GROWTH HORMONES

Reducing the use of antibiotics and artificial growth hormones in the livestock business is in many respects the toughest nut to crack, something that is exceedingly difficult to accomplish without regulatory intervention. In substantial measure, this is because when compared to other facets of the agricultural sector of the economy, the dynamic is different as a result of the demise of the diversified farm and the ascendency of animal factory farms.

Soil conservation practices such as no-till and low-till cultivation are motivated, at least in part, by farmers' love of their land. A substantial majority of farmers, though not all, care about their land and believe in practicing good stewardship of their land so that it is still there for future generations. As noted, Sally Puttmann, who along with her husband operated a diversified farm in Iowa on which they raised corn, soybeans, hogs, and cattle long after most other farmers opted for specialization, characterized sustainable agriculture as agricultural practices "that will not only economically sustain the family that farms the land, but that will enhance the land long after [that] family is gone."[43] Underlying this view is both a sense of stewardship of the land and a sense of obligation to future generations.

One is hard-pressed to find similar sentiments expressed by the owners of the large animal factory farms that produce the bulk of the meat and eggs that come to market in the United States and some other countries. Many of these operations are owned, not by farmers, but by investors with no connection with the land, investors who are interested in only one thing—making as much money as they can. Leopold's land ethic places strong emphasis on the importance of love for the land. Where there is love for the land, stewardship and sustainable agriculture practices follow. Where there is only love of money, farm land and farm animals suffer.

Yet, there are at least a few rays of hope. There are some farmers who have chosen to discontinue nontherapeutic use of antibiotics in the production of livestock. A statement signed by more than 50 farmers and ranchers in the United States notes, "We believe the imprudent use of antibiotics not only renders antibiotics less effective for sick animals, it also threatens and safety of our nation's food supply . . . . We are concerned about the health of our customers, our neighbors, our employees, and our own families." Those signing the document state that the success of their farms and ranches proves that it is "not only possible but actually economically viable to produce meat, dairy products, and eggs that are safe to eat without continually dosing animals with drugs they don't need."[44]

In an article entitled "Antibiotic Use in Animal Agriculture: What Have We Learned and Where Are We Going?" Scott a McEwen, who is affiliated with Ontario Veterinary College in Canada, notes, "Successful animal health management is rarely based on a single practice or technology, including antibiotics. Age-segregation [of animals], all-in all-out management [keeping a group of animals in the same space until they are ready to be sent to market, with no new animals added], biosecurity, sanitation, and vaccination are just a few examples of nonantibiotic practices." He adds, however, "So far, none of these measures is sufficient to completely eliminate the need for antibiotics on farms . . . ."[45] He further notes, "Much more action is needed to ensure that antibiotic use in animals is consistent with prudent or judicious use principles."[46]

The Pew Commission's *Report on Industrial Farm Production* observes, "Alternatives to the present concentrated animal feeding operation (CAFO) model of raising hogs vary widely by region. But although they differ in design, alternative systems share one common element: they all increase both labor and animal husbandry required to manage the animals."[47] In short, they increase costs of production when compared to concentrated animal feeding operations.

Consumer concern about nontherapeutic use of antibiotics and the use of artificial growth hormones is having some impact. It is not unusual to see on bottles of milk statements such as "Our farmers pledge not to treat their cows with artificial growth hormones," though, it should be added that without a certification process similar to that for organic farming, it is difficult to determine whether what is pledged is implemented in practice. McDonald's, which sells billions of hamburgers in countries throughout the world, has adopted policies that favor suppliers who comply with relevant principles for sustainable use of antibiotics. The company's 2003 global antibiotic policy states, "McDonald's prohibits the use of antibiotics belonging to classes of compounds approved for use in human medicine when used solely for growth promotion purposes."[48] What McDonald's envisions is specified in greater detail in a 2015 statement entitled "McDonald's Global Vision for Antimicrobial Stewardship in Food Animals," which prohibits "the use of antimicrobials [antibiotics] in food animals that are by WHO [World Health Organization] definition 'critically important' to human medicine . . . ."[49] Chick-fil-A announced plans to phase out the sale of chicken raised with antibiotics by 2019.[50] A 2017 survey by a consortium of consumer organizations, while recognizing the progress that McDonald's and Chick-fil-A have made, gives the highest marks to Panera Bread and Chipotle Mexican Grill, with many other restaurants and fast food outlets receiving failing grades.[51]

The U.S. Food and Drug Administration (FDA) has taken steps to limit the use of antibiotics in food-producing animals. In April of 2012, the (FDA) banned some uses of antibiotics.[52] That was followed by the Veterinary Feed Directive (VFD), which moved a number of antimicrobial drugs from being available over-the-counter (OTC), to VFD status, effective January 1, 2017.[53] This measure requires that VFD-listed drugs be available only "within the context of a veterinarian-client-patient relationship."[54]

Another layer of protection is provided by the National Antimicrobial Resistance Monitoring System (NARMS), a joint effort of several government agencies. As part of this program, the U.S. Department of Agriculture (USDA) checks for antibiotic residues in meat. Meat with residues of concern does not make it to the market.[55]

In short, progress has been made. These are all steps in the right direction. Whether these measures prove adequate remains to be seen.[56]

HOW SAFE IS ROUNDUP (GLYPHOSATE)?

There is heated debate as to how safe glyphosate is. A press release issued a few years ago by the Organic Consumers Association states that there is a link between glyphosate and cancer.[57] The news release references a study in Sweden done by Lennart Hardwell and Mikael Eriksson reported in the 15 March 1999 edition of the *Journal of the American Cancer Society*, which suggested that there might be a link between exposure to herbicides and the incidence of non-Hodgkin lymphoma (NHL). Though much of the study pertains to herbicides and pesticides other than glyphosate, Hardwell and Eriksson state that glyphosate "also might be of concern." They conclude that "glyphosate deserves further epidemiologic studies."[58]

Others argue that glyphosate poses no risk to human health. In an article published in *Regulatory Toxicology and Pharmacology* (an on-line journal) in 2000, Gary M. Williams, Robert Kroes, and Ian C. Munroe state that, based on their review of various studies, "glyphosate is noncarcinogenic." They add that "under present and expected conditions of new use, there is no potential for Roundup herbicide to pose a health risk to humans."[59] (Monsanto, the company that produces Roundup, gave the authors access to toxicological information, with company employees providing scientific support.)

Laboratory studies indicate that caffeine, which is found in coffee and many soft drinks, is far more toxic than glyphosate. One commentator observes, "Should this be cause for concern . . . ? No, the main reason being that a typical dosage of caffeine is not high enough to cause toxicity." A 140 lb. person would need to drink 128 cups of coffee, one after the other, for the caffeine to cause toxicity problems.[60] Not even the most avid coffee drinkers consume that much coffee! And what about Roundup? A person would have to drink about three gallons of Roundup for it to be toxic.[61] That is not going to happen.

But might there be long-term adverse health effects resulting from exposure to glyphosate? Critics insist that there can be. One frequently-referenced article states that consequences of exposure to glyphosate include "most of the diseases and conditions associated with a Western diet, which include gastrointestinal disorders, obesity, diabetes, heart disease, depression, autism, infertility, cancer and Alzheimer's disease."[62]

These claims, however, are sharply disputed. For example, in a review of epidemiologic studies of glyphosate and cancer published in *Regulatory Toxicology and Pharmacology*, Pamela J. Mink et al. state, "Our review of the currently available epidemiologic literature on glyphosate and cancer found no evidence of a consistent pattern of positive associations that would be indicative of a causal relationship between any site-specific cancer and exposure to glyphosate."[63]

During the Obama administration, the U.S. Environmental Protection Agency (EPA) fact sheet on glyphosate stated, "EPA's worst case risk assessment of glyphosate's many registered food uses concludes that human dietary exposure and risk are minimal . . . . Exposure to workers and other applicators is not expected to pose undue risks, due to glyphosate's low acute toxicity. However, splashes during mixing and loading of some products can cause injury, primarily eye and skin irritation. EPA is continuing to recommend [personal protective equipment], including protective eye wear . . ."[64]

The EPA position on this matter is not universally applauded. Some believe that political considerations play a role in issuing guidelines with respect to glyphosate and other agricultural chemicals—criticism which gained momentum in March of 2015 when the World Health Organization International Agency for Research on Cancer (IARC) classified glyphosate as "probably carcinogenic to humans."[65] In the wake of the IARC report, the California Office of Environmental Health Hazard Assessment said in September of 2015 that it planned to add glyphosate to its list of known carcinogens. Monsanto disputed this assessment and responded by suing the California Office of Environmental Health Hazard Assessment and its acting director, citing numerous studies indicating that glyphosate is safe.[66]

In 2016, a joint meeting of the Food and Agriculture Organization of the United Nations (FAO) Panel of Experts on Pesticide Residues in Food and the Environment and the World Health Organization (WHO) Core Assessment Group on Pesticide Residues (JMPR) was held at WHO headquarters in Geneva. Following the meeting, the FAO and WHO issued a report stating that "glyphosate is unlikely to pose a carcinogenic risk to humans from exposure through diet."[67] The statement did not address the question of whether skin contact or inhaling glyphosate might be carcinogenic.

In November of 2017, the European Union (EU) voted to extend its approval for the use of glyphosate for five years. Eighteen member states voted in favor of the proposal while nine voted against the proposal and one abstained. The vote was weighted by population size.[68]

On December 13, 2017, the U.S. Environmental Protection Agency released for public comment a draft of a statement on risk assessments for glyphosate, which concluded that glyphosate is not likely to be carcinogenic to humans. The statement notes, "EPA's human health review evaluated dietary, residential/non-occupational, aggregate and occupational exposures. Additionally, the Agency performed an in-depth review of the glyphosate cancer database, including data from epidemiological, animal carcinogenicity, and genotoxicity studies."[69]

However, eight months later a jury in a state court in San Francisco ruled in favor of Dewayne Johnson, a school groundskeeper whose lawyers argued that Roundup made by Monsanto (which is now owned by Bayer[70]) likely caused his cancer (he was diagnosed with non-Hodgkin's lymphoma). The court awarded Johnson $289 million in damages.[71] A California judge, while upholding the jury's verdict, reduced the amount to be paid to $78 million.[72]

A second case involving Roundup, this one in a federal court in San Francisco, was also decided in favor of the plaintiff. Edwin Hardemann had used Roundup to control weeds and poison oak on his property for twenty-six years. In 2015 he learned that he had non-Hodgkin's lymphoma. The jury awarded him $80 million, which included $75 million for punitive damages.[73]

A third case, this one in a state court in Alameda County in California, resulted in an even larger verdict against Monsanto. The jury ordered Monsanto to pay Alva and Alberta Pilliod $2 billion in damages. They had used Roundup on their Northern California property for decades. Both developed non-Hodgkin's lymphoma.[74]

Whether these court decisions will be upheld on appeal remains to be seen. What is clear, however, is that assessing possible health risks related to glyphosate is hampered by a paucity of independent scientific studies on the health impact of glyphosate. This is an area in which more research is urgently needed. Hardwell and Eriksson were right when they stated that "glyphosate deserves further epidemiologic studies."

A GLASS HALF FULL OR A GLASS HALF EMPTY?

In a poem included in *Clearing*, Wendell Berry observes:

> A man who does not ask too much
> becomes the promise of his land.[75]

Farmers who take good care of their land by practicing sustainable agriculture are the promise of their land. They exemplify civic virtue in action, contributing to the common good, not just of those who now inhabit planet earth but of future generations as well.

At the same time, much remains to be done. One only needs to drive through the hilly farmland of western Iowa during springtime to gain a sense of that. Some fields are scarred by gullies where spring rainstorms have washed away valuable topsoil. In other fields, grass and bushes protect the soil in drainage areas from erosion. A glass half full or a glass half empty? Perhaps both.

SUMMING IT ALL UP

The productivity revolution in agriculture—and the environmental damage caused by it—almost defy comprehension. The good news, however, is that the sustainable agriculture movement, motivated in substantial measure by the love for the land of which Leopold spoke, has substantially changed the nature of agriculture, not only in the United States but in many other countries as well. Granted, much remains to be done. Difficult challenges lie ahead. Yet we can move toward the future with quiet confidence—quiet confidence born of the realization that much can be accomplished by individuals and communities working together to practice stewardship with respect to that with which we have been entrusted.

NOTES

1. Food and Agriculture Organization (FAO) of the United Nations, "Sustainable Agriculture and Rural Development" at http://www.fao.org/docrep/U8480E/U8480E0l.htm (last accessed January 15, 2019).
2. U.S. Department of Agriculture National Institute of Food and Agriculture, "Legal Definition of Sustainable Agriculture" at https://nifa.usda.gov/program/sustainable-agriculture-program (last accessed June 12, 2019).
3. Agriculture Sustainability Institute of the University of California—Davis, "What Is Sustainable Agriculture?" at http://www.sarep.ucdavis.edu/ (last accessed January 17, 2013).

4. Quoted in Daniel E. Lee, *Generations and the Challenge of Justice* (Lanham, MD: University Press of America, 1996), 108.

5. This is not a matter of complete consensus. Some suggest that organic farming leaves a larger carbon footprint and does less to protect the soil from erosion than do some other forms of agriculture, matters that will be discussed later in this chapter. (I am indebted to Chet Boruff for these observations. A farmer, he is the CEO of the Association of Official Seed Certifying Agencies.)

6. *Organic Foods Production Act of 1990* at https://www.ams.usda.gov/sites/default/files/media/NOPOFPA.pdf (last accessed June 12, 2019). See also "Introduction to Organic Practices," United States Department of Agriculture at https://www.ams.usda.gov/sites/default/files/media/Organic%20Practices%20Factsheet.pdf (last accessed June 12, 2019).

7. United States Department of Agriculture, "Organic Regulations" at https://www.ams.usda.gov/rules-regulations/organic (last accessed June 12, 2019).

8. See, for example, European Commission, "Inspection and Certification—Organic Farming" at http://ec.europa.eu/agriculture/organic/consumer-confidence/inspection-certification_en (last accessed June 12, 2019).

9. "U.S. Organic Sales Break through $50 Billion Mark in 2018," Organic Trade Association at https://ota.com/news/press-releases/20699 (last accessed June 16, 2019).

10. These prices were in effect at the Jewell Store at Cityline Plaza in Moline, Illinois, on December 17, 2018.

11. Government subsidies could offset the higher costs of production for organic produce, grains, eggs, and meat. However, with the United States, EU countries and other countries struggling to reduce substantial budget deficits, this is not likely to be the wave of the future.

12. U.S. Environmental Protection Agency, "Integrated Pest Management" at http://www.epa.gov/agriculture/tipm.html (last accessed January 29, 2013).

13. University of Illinois Extension, "Integrated Pest Management" at http://www.extension.org/pages/17849/integrated-pest-management (last accessed January 29, 2013).

14. "Global Positioning System (GPS): Space Segment," GPS.Gov: Official U.S. Government Information about the Global Positioning System (GPS) and Related Topics at http://www.gps.gov/systems/gps/space/ (last accessed January 15, 2019).

15. When the U.S. Department of Defense first introduced GPS, the signals were partially encrypted to make it difficult for unfriendly countries to use U.S. satellites to target their missiles. When the Cold War ended, the encryption was phased out, allowing anyone anywhere in the world with the appropriate equipment to use GPS signals to determine location. Sophisticated GPS systems, such as those used by many land surveyors, enable determining locations within a few inches. (See, e.g., "Land Surveying GPS" at http://www.gps.gov/systems/gps/space/ [last accessed January 15, 2019].)

16. "GPS.Gov: Agricultural Applications" at http://www.gps.gov/applications/agriculture/ (last accessed June 12, 2019).

17. Ibid.

18. John Fraser Hart, "Change in the Corn Belt."

19. Another widely used agricultural herbicides that degrades quite rapidly is 2,4-D, which kills broadleaf plants. For additional information on 2,4-D, see National Pesticide Information Center, "2,4-D Technical Fact Sheet" at http://npic.orst.edu/factsheets/2,4-DTech.pdf (last accessed June 12, 2019).

20. National Pesticide Information Center (NPIC), "Glyphosate Technical Fact Sheet" at http://npic.orst.edu/ingred/glyphosate.html (last accessed June 12, 2019).

21. Ibid.

22. Extension Toxicology Network, "Glyphosate" athttp://pmep.cce.cornell.edu/profiles/extoxnet/dienochlor-glyphosate/glyphosate-ext.html (last accessed June 12, 2019).

23. Pestizid Aktions Netzwerk (PAN), "Phasing in Alternatives to DDT" at http://www.pan-germany.org/download/ddt/phasing_in_alternatives_to_ddt.pdf (last accessed June 12, 2019).

24. *The Stockholm Convention on Persistent Organic Pollutants* at http://chm.pops.int/Convention/ConventionText/tabid/2232/Default.aspx (last accessed June 12, 2019).

25. Joanne Silberner, "WHO Backs Use of DDT Against Malaria," National Public Radio, September 15, 2006, at http://www.npr.org/templates/story/story.php?storyId=6083944 (last accessed June 12, 2019).

26. PAN, "Phasing in Alternatives to DDT."

27. Illinois Department of Public Health, "Pyrethroid Insecticides" at http://www.idph.state.il.us/envhealth/factsheets/pyrethroid.htm (last accessed June 12, 2019).

28. "Dimethoate" at http://pan-uk.org/pestnews/Actives/dimethoa.htm (last accessed February 1, 2013). As this article also notes, however, dimethoate is not devoid of controversy. Concern has been expressed about health risks experienced by those exposed to it during manufacture and use. Concern has also been expressed about the environmental impact of neonicotinoid insecticides which, critics charge, adversely affect bees. The European Union has asked members states to restrict their use in order to protect bees ("EU Seeks to Protect Bees from Agricultural Insecticides," *Europe Online Magazine*, January 31, 2013, at http://en.europeonline-magazine.eu/eu-seeks-to-protect-bees-from-agricultural-insecticides_261829.html [last accessed June 12, 2019]).

29. Rolf Derpsch, Theodor Friedrich, Amir Kassam, and Li Hongwen, "Current Status of Adoption of No-Till Farming in the World and Some of Its Main Benefits," *International Journal of Agricultural and Biological Engineering* (March 2010) at http://www.fao.org/ag/ca/CA-Publications/China_IJABE.pdf (last accessed June 12, 2019).

30. David R. Huggins and John P. Reganold, "No-Till: the Quiet Revolution," *Scientific American* (June 30, 2008), 77.

31. Brad Plumer, "No-Till Farming Is on the Rise. That's Actually a Big Deal," *Washington Post*, November 9, 2013.

32. Ibid., 70.

33. Northern Virginia Soil and Water Conservation District, "No-Till Agriculture: Good for the Soil and the Bottom Line," *Conservation Currents* (September 2005) at http://www.fairfaxcounty.gov/nvswcd/newsletter/notill.htm (last accessed June 12, 2019).

34. Edward Bradley, "No Till Farming" at http://www.notillfarming.co.uk/index.html (last accessed June 12, 2019).

35. Huggins and Reganold, "No-Till: The Quiet Revolution," 74.

36. Ibid.

37. Julia Lawton, "Is Conservation Agriculture Better than Organic Farming?" at http://science.howstuffworks.com/environmental/green-science/conservation-agriculture1.htm (last accessed June 12, 2019).

38. Ibid.

39. The advent of HT varieties is not without its downside. Widespread use of herbicides containing glyphosate has resulted in the emergence of glyphosate-resistant weeds, much as the widespread use of antibiotics has resulted in the emergence of antibiotic-resistant bacteria. It is within the range of the possible, of course, that the emergence of glyphosate-resistant weeds would have occurred without the introduction of HT varieties of corn, soybeans, and cotton, though it is probably the case that the use of HT varieties in no-till farming has resulted in increased use of herbicides containing glyphosate.

40. U.S. Department of Agriculture, "Recent trends in GE Adoption" at https://www.ers.usda.gov/data-products/adoption-of-genetically-engineered-crops-in-the-us/recent-trends-in-ge-adoption.aspx (last accessed June 12, 2019).

41. Food and Agriculture Organization of the United Nations, "Genetically Modified Crops" at http://www.fao.org/3/i2490e/i2490e04d.pdf (last accessed June 12, 2019).

42. USDA, "Recent Trends in GE Adoption." Since "the bugs always fight back," it is entirely possible that mutations of crop-threatening insects will result in Bf varieties losing their effectiveness in fighting off pests, just as mutations of various bacteria have made certain antibiotics useless when fighting bacterial infections. As in the case of the scientists who develop new antibiotics, the scientists creating Bf varieties of various crops are likely to be in a never-ending struggle to develop new Bt varieties in order to stay ahead of crop-threatening insects. For a useful overview of some of the environmental risks that might be related to the introduction of Bt and other genetically-engineered crop plants and ways of counteracting these risks, see Jorge Fernandez-/Cornejo and William D. McBride, "Genetically Engineered Crops

for Pest Management in U.S. Agriculture: Farm Level Effects," U.S. Department of Agriculture Economic Research Service, Report No. 786 (April 2000), 3.

43. Quoted in Daniel E. Lee, *Generations and the Challenge of Justice*, 108.

44. Quoted by Bottemiller in "Farmers and Scientists Call for More Action on Antibiotics in Agriculture."

45. Scott A. McEwen, "Antibiotic Use in Animal Agriculture: What We Have Learned and Where Are We Going?" *Animal Biotechnology* (2006) 17: 243.

46. Ibid., 244.

47. Pew Commission, *Report on Industrial Farm Production*, 33.

48. *McDonald's Global Policy on Antibiotic Use in Food Animals*, at https://www.mcdonalds.com/dam/AboutMcDonalds/Sustainability/Sustainability%20Library/antibiotics_policy.pdf (last accessed June 12, 2019).

49. *McDonald's Global Vision for Antimicrobial Stewardship in Food Animals* at http://worldvet.org/uploads/news/docs/antimicrobial_stewardship_vision.pdf (last accessed June 12, 2019).

50. Elizabeth Landau, "Chick-fil-A To Serve Antibiotic-Free Chicken," *Cable News Network (CNN)*, February 12, 2014 at http://www.cnn.com/2014/02/11/health/chick-fil-a-chicken-antibiotics/ (last accessed June 12, 2019). See also Chick-fil-A, "No Antibiotics Ever: Chick-fil-A's Commitment Charts New Territory" at https://thechickenwire.chick-fil-a.com/Inside-Chick-fil-A/No-Antibiotics-Ever-Chick-fil-As-Commitment-Charts-New-Territory (last accessed June 12, 2019) and Chick-fil-A, "Our Journey: Antibiotic-Free Chicken" at http://www.chick-fil-a.com/Antibiotic-Free (last accessed June 12, 2019).

51. *Chain Reaction III: How Top Restaurants Rate on Reducing Use of Antibiotics in their Meat Supply*, 2017, at https://1bps6437gg8c169i0y1drtgz-wpengine.netdna-ssl.com/wp-content/uploads/2017/09/Chain-Reaction-III.pdf (last accessed June 12, 2019).

52. "FDA To Phase Out Non-Therapeutic Use of Antibiotics in Food Animals," *Medical News Today*, December 12, 2013, at http://www.medicalnewstoday.com/articles/270010.php (last accessed October 23, 2015). See also "FDA Hopes to Curb Antibiotic Use on Farms," *Cable News Network (CNN)*, December 12, 2013, at http://www.cnn.com/2013/12/11/health/fda-antibiotics-farms/ (last accessed October 23, 2015).

53. U.S. Food and Drug Administration, "Drugs Transitioning from Over-the-Counter (OTC) to Veterinary Feed Directive (VFD) Status" at https://www.fda.gov/downloads/AnimalVeterinary/DevelopmentApprovalProcess/UCM482258.pdf (last accessed June 12, 2019).

54. U.S. Food and Drug Administration, "Fact Sheet: Veterinary Feed Directive Final Rule and Next Steps" at https://www.fda.gov/AnimalVeterinary/DevelopmentApprovalProcess/ucm449019.htm (last accessed June 8, 2018). See also U.S. Food and Drug Administration, "Veterinary Feed Directive (VFC)" at https://www.fda.gov/AnimalVeterinary/DevelopmentApprovalProcess/ucm071807.htm (last accessed June 12, 2019).

55. U.S. Food and Drug Administration, "About NARMS" at https://www.fda.gov/AnimalVeterinary/SafetyHealth/AntimicrobialResistance/NationalAntimicrobialResistanceMonitoringSystem/ucm059089.htm (last accessed June 12, 2019).

56. For a useful overview of the problem, see Timothy F. Landers, Bevin Cohen, Thomas E. Wittum, and Elaine L. Larson, "A Review of Antibiotic Use in Food Animals: Perspective, Policy, and Potential," *Public Health Reports,* January–February 2012, 4–22, at http://www.ncbi.nlm.nih.gov/pmc/articles/PMC3234384/ (last accessed June 12, 2019). They conclude, "It is now critical that agricultural use of antibiotics be recognized as one of the major contributors to the development of resistant organisms that result in life-threatening human infections and included as part of the strategy to control the mounting public health crisis of antibiotic resistance."

57. Organic Consumers Association, "New Study Links Monsanto's Roundup to Cancer," June 22, 1999, at https://www.organicconsumers.org/old_articles/Monsanto/glyphocancer.php (last accessed March 24, 2016).

58. Lennart Hardwell and Mikael Erikson, "A Case-Controlled Study of Non-Hodgkin Lymphona and Exposure to Pesticides," *Journal of the American Cancer Society* (March 15, 1999) 85:1353–1300.

59. Gary M. Williams, Robert Kroes, and Ian C. Munroe, "Safety Evaluation and Risk Assessment of Herbicide Roundup and Its Active Ingredient, Glyphosate, for Humans," *Regulatory Toxicology and Pharmacology* (2000), 31:117–165 at http://www.msal.gob.ar/agroquimicos/pdf/Williams-et-al-2000.pdf (last accessed September 10, 2016).

60. Xiao Lim, "Is Glyphosate, Used with Some GM crops, Dangerously Toxic to Humans?" *Genetic Literacy Project*, April 30, 2014, at https://www.geneticliteracyproject.org/2014/04/30/is-glyphosate-used-with-some-gm-crops-dangerously-toxic-to-humans/ (last accessed June 12, 2019).

61. Ibid.

62. Anthony Samsel and Stephanie Seneff, "Glyphosate's Suppression of Cytochrome P450 Enzymes and Amino Acid Biosynthesis by the Gut Microbe: Pathways to Modern Diseases," *Entropy* [2003] 15:1416. See also Siriporn Thongprakaisang et al., "Glyphosate Induces Human Breast Cancer Cells Growth via Estrogen Receptors, *Food and Chemical Toxicology* (2013) 59:129–136, and Celine Gasnier et al., "Glyphosate-Based Herbicides Are Toxic and Endocrine Disruptors in Human Cell Lines," *Toxicology* (2009) 262:184–191.

63. Pamela J. Mink, Jack S. Mandel, Bonnielin K Sceuman, and Jessica I. Lundin, "Epidemiologic Studies of Glyphosate and Cancer: A Review," *Regulatory Toxicology and Pharmacology* (2012), 63: 440–452 at http://www.sciencedirect.com/science/article/pii/S0273230012000943 (last accessed September 10, 2016). A conflict of interest statement at the end of the article states that Mandel "has served as a paid consultant to Monsanto Company. Final decisions regarding the content of the manuscript were made solely by the four authors." The acknowledgements note that "research was supported by Monsanto Company, St. Louis, Missouri."

64. U.S. Environmental Protection Agency, "Glyphosate" at http://nepis.epa.gov/Exe/ZyNET.exe?Client=EPA&Index=1991+Thru+1994&Docs=&Query=%28PPE%29+or+%28PPE%29+OR+FNAME%3D%22200005X7.txt%22+AND+FNAME%3D%22200005X7.txt%22&FuzzyDegree=0&ZyAction=ZyActionD&User=ANONYMOUS&Password=anonymous&SortMethod=h%7C-&MaximumDocuments=1&ImageQuality=r75g8%2Fr75g8%2Fx150y150g16%2Fi425&Display=p%7Cf&DefSeekPage=x&Time=&EndTime=&SearchMethod=1&Toc=&TocEntry=&TocRestrict=n&QField=&QFieldYear=&QFieldMonth=&QFieldDay=&UseQField=&IntQFieldOp=0&ExtQFieldOp=0&XmlQuery=&SearchBack=ZyActionL&Back=ZyActionS&BackDesc=Results+page&MaximumPages=1&ZyEntry=&SeekPage=&ZyActionID=&File=D%3A%5C%5CZYFILES%5C%5CINDEX+DATA%5C%5C91THRU94%5C%5CTXT%5C%5C00000007%5C%5C200005X7.txt&Doc=%3Cdocument+name%3D%22200005X7.txt%22+path%3D%22D%3A%5CZYFILES%5CINDEX+DATA%5C91THRU94%5CTXT%5C00000007%5C%22+index%3D%221991+Thru+1994%22%2F%3E&QueryTerms=PPE [last accessed March 24, 2016]).

65. Reuters, "W.H.O. Report Links Ingredient in Roundup to Cancer," *New York Times*, March 20, 2015.

66. Reuters, "Monsanto Sues California over Herbicide Classification," *New York Times*, January 21, 2016.

67. "Joint FAO/WHO Meeting on Pesticide Residues Summary Report," May 16, 2016, at https://www.who.int/foodsafety/jmprsummary2016.pdf (last accessed January 15, 2019). See also, "Food Safety: Frequently Asked Questions," World Health Organization, May 27, 2016, at https://www.who.int/foodsafety/faq/en/ (last accessed January 15, 2019).

68. Danny Hakim, "Glyphosate, Top-Selling Weed Killer, Wins E.U. Approval for 5 Years," *New York Times*, November 27, 2017.

69. United States Environmental Protection Agency, "EPA Releases Draft Risk Assessments for Glyphosate" at https://www.epa.gov/pesticides/epa-releases-draft-risk-assessments-glyphosate (last accessed June 8, 2018). See also U.S. Environmental Protection Agency, "Draft Human Health and Ecological Risk Assessments for Glyphosate" at https://www.epa.gov/ingredients-used-pesticide-products/draft-human-health-and-ecological-risk-assessments-glyphosate (last accessed June 8, 2018).

70. Bayer, a German conglomerate, purchased Monsanto in June of 2018.

71. "Jury Orders Monsanto to Pay Nearly $290 Million in Roundup Trial," *NBC News*, August 11, 2018 at https://www.nbcnews.com/news/us-news/jury-orders-monsanto-pay-290m-roundup-trial-n899811 (last accessed December 18, 2018). See also Sam Levin, "The Man Who Beat Monsanto: 'They Have To Pay for Not Being Honest,'" *The Guardian*, September 26, 2018.

72. "Judge Upholds Monsanto Verdict but Cuts Award to $78 Million," Associated Press, October 22, 2018.

73. Julia Jacobs, "Monsanto Ordered to Pay $80 Million in Roundup Cancer Case," *New York Times*, March 27, 2019. See also Mihir Zaveri, "Monsanto Weedkiller Roundup Was 'Substantial Factor' in Causing Man's Cancer, Jury Says," *New York Times*, March 19, 2019.

74. Patricia Cohen, "$2 Billion Verdict Against Monsanto Is Third to Find Roundup Caused Cancer," *New York Times*, May 13, 2019.

75. Wendell Berry, "The Clearing" in *Clearing* (New York: Harcourt Brace Jovanovich, 1977), 24.

Part III

Recycling

Chapter Six

Recycling Across the Ages

Recycling is nothing new. In fact, it has been around for centuries. Take, for example, the Arch of Constantine, located in Rome between the Colosseum, which in ancient Rome was the site of bloody gladiatorial contests, and Palatine Hill, where the lavish homes of emperors once stood. Dedicated in 315, the arch was erected to commemorate Constantine's victory over Maxentius at the Battle of Milvian Bridge in 312, a victory which brought Constantine to the pinnacle of power in fourth-century Rome. The architects of the arch reused several reliefs and statues from second-century imperial monuments. Flanking the inscription on the central part of the monument on both sides are pairs of panels taken from a monument honoring Marcus Aurelius, eight panels in all. (Three additional panels, which appear to be of the same vintage, can be seen in the Capitoline Museum, which overlooks the Roman Forum.) Above each of the lower arches on both sides of the monument (the Arch of Constantine features a large central arch and two lower side arches) are two round medallions depicting various events in the life of Hadrian, with Hadrian's facial features reworked to make them appear more like those of Constantine. The architects of the Arch of Constantine also repurposed statues from the time of Trajan, mounting them on the upper level of the arch adjacent to the panels taken from the monument honoring Marcus Aurelius.[1]

The practice of architectural recycling did not stop with the Arch of Constantine. In fact, it was only the beginning with worse yet to come. It is widely assumed that the buildings and monuments of ancient Rome were reduced to rubble by invading barbarians. For the most part, that is not true. The prime culprits were architects who viewed buildings and monuments from ancient Rome as useful sources of marble and other materials for their own construction projects. This caused far more damage to the once-elegant buildings and monuments than anything the Visigoths, the Vandals, and other groups of invading so-called barbarians ever did. In the fourth and fifth centuries, emperors repeatedly issued laws prohibiting the stripping of marble and other materials from buildings and monuments, but with little effect.[2]

Even buildings such as the Pantheon which survived relatively intact (the Pantheon, rebuilt by Hadrian in 125, is still in use today) did not escape the fate of having materials used in their construction taken for other purposes. Urban VIII, a member of the Barberini family who served as pope from 1623–1644, authorized removing the bronze from the ceiling of the portico of the Pantheon to melt down for cannon and to make the spiral columns of the *baldacchino* (canopy) over the high altar in St. Peter's Basilica, a Baroque masterpiece that is the work of Gian Lorenzo Bernini. This prompted an anonymous wag to write, "Quod non fecerunt barbari, fecerunt Barberini (What the barbarians did not do, the Barberini did)."

REUSE OF OLD CLOTHING AND OTHER TEXTILES

In the course of the centuries, recycling has taken many twists and turns, manifesting itself in many different ways and experiencing both ebb and flow, with, in time, architectural recycling becoming little more than a historical footnote as other forms of recycling gained prominence. The reuse of old clothes gave rise to a second-hand clothes trade in Europe and elsewhere. In some cases, this was the result of discriminatory measures imposed on segments of the population. In 1555, Pope Paul IV, one of the most mean-spirited individuals ever to occupy the Throne of St. Peter, issued a bull entitled *Cum nimis absurdum* which subjected Jews living in the papal states[3] to harsh measures. One of the provisions of this bull stated that "Jews may carry on no business as purveyors of grain, barley, or other items necessary for human sustenance, but must be limited [in this sphere] to dealing only in second-hand clothing . . . ."[4] *Museo Ebraico di Roma* (the Jewish Museum of Rome) includes in its collection numerous beautiful *Sefer Torah* mantles (covers for Torah scrolls) and other items made from old drapes and other recycled textiles.[5]

In other cases, the market for used clothing and cloth was driven by economic factors, be they thrift in the case of those purchasing used clothing and cloth, or economic opportunity in the case of those selling them. Take, for example, Renaissance Florence, which was a center of the textile and garment industry. Prior to the Industrial Revolution, high-quality new cloth was a luxury that could be afforded only by the wealthy, putting clothing made of fine cloth beyond the reach of those who were less affluent. Used-clothing and used-cloth dealers (*rigattieri*) served this market. Even members of wealthy families would sometimes have clothing made of used cloth (*panno usato*), which they always distinguished from old cloth (*panno vec-*

chio) or worn-out cloth (*panno triste*).[6] In fifteenth-century Florence, there were formally-established trading guilds that offered for sale second-hand silk gowns and items made of previously-used velvet. Lower-class itinerant vendors, many of them women, worked the streets buying and selling used clothes of any quality from households of different classes.[7]

It was much the same elsewhere. In a book chapter entitled "Shifting Currency: The Culture and Economy of the Second Hand Trade in England, c. 1600–1850," Beverly Lemire notes, "In the long period of human history in the West before the full force of industrialization changed the world, the scarcity of goods ensured their careful husbanding. Use and reuse defined the everyday for all but a tiny minority of the population."[8] She adds, "London, like other great cities of Europe, was especially rich in used garments of every quality . . . ."[9]

The recycling of used clothing did not always involve market transactions. Much of it happened at home as seamstresses repaired and remade clothing for members of their families or for those for whom they worked. This included families who were more affluent, as well as those who were not. In *Waste and Want*, Susan Strasser notes:

> Repairing garments and linens often went far beyond mending to "making over." The term covered a lot of territory: simple hemming, dyeing or treating worn fabrics, covering frayed cuffs and collars with handmade needlework or machine-made braiding, or completely disassembling a garment and using the pieces for some other purpose.[10]

SCRAP METAL

Recycling, of course, has historically involved far more than wearing second-hand clothing. Boston silversmith Paul Revere, immortalized in Henry Wadsworth Longfellow's epic poem, "Paul Revere's Ride,"[11] saved scrap metal for reuse, as did many others[12]—a practice that continued well into the twentieth century. In many cases, craftsmen such as Revere saved scrap metal for their own use, much as farmers for generations have saved worn-out farm machinery to provide wheels, axles, and other parts for other projects. In time, however, scrap dealers became the middlemen, purchasing scrap metal wherever they could find it and reselling it to foundries to melt down into ingots, with the foundries in turn selling these ingots to factories and others who could use it as a raw material for whatever they were making.[13]

RAGS FOR THE PAPER INDUSTRY

Scrap metal was not the only commodity for which there was a market. When clothing was so worn that is was no longer saleable as used clothing, it was a valuable source of rags for the paper industry, as were badly-worn tablecloths and other textiles. Both linen and cotton were in considerable demand with linen particularly prized because it could be used to make paper of high quality. Using worn-out linen tablecloths and cotton garments to make paper was less labor intensive and hence less expensive than raising flax and cotton and milling them to make virgin linen and cotton paper. Moreover, the quality of the paper made from recycled textiles is often equal or superior to that made from virgin materials.[14] By 1850, the U.S. demand for rags to be used to manufacture paper had grown to the point that rags were imported from Europe and elsewhere. That year, 98 million pounds of rags were imported with the total increasing to 123 million pounds twenty-five years later. Most of the imported rags came from the United Kingdom and British possessions, with the rest coming from Italy, Austria, Turkey, Germany, and about twenty other countries.[15]

REFUSE COLLECTORS AND DUST YARDS

During the first three-quarters of the nineteenth century, recycling of household wastes was quite common in England and elsewhere. In many urban areas, household refuse was collected by scavengers and dustmen and taken to dust yards, where those who worked there, mainly women paid meager wages, would rummage through the refuse looking for reusable items such as brass, rags, and waste paper. Though the women who sorted through the household refuse at dust yards were paid subsistence wages (or less), some enterprising individuals managed to become wealthy in the refuse business. Among them was Henry Dodd, who became known as the "Golden Dustman" as a result of the wealth he accumulated in the refuse business and other enterprises. Born in 1801, Dodd worked as a ploughboy in fields within sight of St. Paul's Cathedral in London. In time, he became a refuse collector (scavenger), which he developed into a highly successful business. His success in the refuse business, brick making, and other enterprises made him one of the wealthiest individuals in London at the time of his death in 1881.[16]

As the nineteenth century gave way to the twentieth century, many of the dust yards were replaced by incinerators which burned refuse. Incinerators were believed to be more hygienic and a more complete method of disposal than was recycling, notwithstanding the fact that the incinerators emitted noxious smoke and fumes and left clinkers and ashes that had to be dumped somewhere.[17]

WARTIME RECYCLING

World War I brought renewed interest in recycling—renewed interest born of necessity. For centuries, bells had been melted down to make cannon[18] and statues to make bullets during times of war.[19] However, World War I, which was known as the Great War in Europe at the time because no one knew that in the years to come there would be an even more destructive war, ushered in a new chapter in the history of wartime recycling.

The senseless war in Europe was a bloody affair that snuffed out the lives of millions at the Marne and at the Somme, at Ypres and at Verdun, and at countless other places both large and small where opposing armies fought bitter battles for little gain. It had begun as a war of maneuver but soon bogged down in the stalemate of trench warfare, with zigzagging and interlocking trenches stretching from the Swiss border to the English Channel, trenches that were muddy and miserable when the weather was bad and infested with rats regardless of whether the weather was good or bad. With the introduction of rapid-firing long-range artillery, machine guns, and other weapons of mass destruction, military tactics that were refined during the Napoleonic era no longer worked. From the safety of headquarters in the rear far beyond the range of enemy artillery, unimaginative and unscrupulous generals, who were quite willing to sacrifice the lives of others, ordered assault after assault, notwithstanding the fact that the lethal combination of barbed wire, which made it difficult to advance, and machine guns, which could fire hundreds of rounds a minute, made "going over the top" into the no-man's land separating the trenches of the opposing armies a virtual death sentence.

Warfare on this scale in an industrialized age consumed massive amounts of ammunition and supplies while calling for the production of huge numbers of rifles, machine guns, cannons, and other weapons of war. With many essential materials and goods in short supply, there came to be a premium on salvaging and recycling anything useful to the war effort.

Some of this was economically driven. The interruption of trade after 1914 resulted in increased prices for recovered waste materials, creating a temporary boom in the waste recovery and scrap trades. With an eye on the profits to be made from waste, many local authorities in England and elsewhere established collection centers for the recovery of rags, ferrous and non-ferrous metals, and organic matter from household refuse.[20]

World War I was followed by a decade of indulgence. Then in October of 1929, the New York stock market crashed, with losses for the month totaling $16 billion, an astronomical amount of money in those days.[21] The world was changed forever. Almost overnight, the giddy days of the earlier years of the decade were but distant memories. By 1933, the U.S. gross domestic product (GDP) had shrunk by nearly 50 percent while the unemployment rate had climbed to 24.9 percent of the workforce.[22] As the Great Depression spread across the Atlantic, millions more were without work in the United Kingdom, France, and elsewhere. As had been the case in previous eras of hard times, frugality became a necessity and everything that could be reused was reused.

On September 1, 1939, Adolf Hitler sent his panzer divisions into Poland, igniting a war that few wanted—a war that in time was to involve most nations of the world. A war that claimed almost four times as many lives as the horrific bloodbath of 1914–1918.

After the fall of France, the United Kingdom stood virtually alone against the Nazi juggernaut, its all-important lifeline to America under constant attack by German U-boats. In the first month of the war, twenty-four British merchant ships were torpedoed, with six more lost to gunfire and mines.[23] That was only the beginning. By the time the war ended nearly six years later, 2,426 British merchant ships had been destroyed by enemy action."[24] British Prime Minister Winston Churchill later recalled that "the only thing that ever really frightened me was the U-boat peril."[25]

During World War II, the disruption to the United Kingdom's international supply lines was far greater than had been the case in the previous war, which made salvage measures of utmost importance. The government created a salvage department that was part of the Ministry of Supply to encourage salvaging on the local level. Initially the Ministry of Supply relied on exhortation in an effort to encourage recycling, but when it became apparent that many local authorities were ignoring the call to salvage, the government resorted to compulsory measures. From October 1939 through July 1947, salvage programs recovered 2.1 million tons of waste paper, 1.8 million tons of scrap metals, and 2.4 million tons of domestic kitchen waste[26] A Scottish newspaper observed, "In war-time, it is necessary to exercise the strictest economy in the use of our resources, and the careful collection of waste materials is a humble, but useful, way of contributing to the national effort."[27]

After the Japanese attack on Pearl Harbor, the United States declared war on Japan—but not on Germany. Hitler then made one of his greatest blunders by declaring war on the United States, which responded by declaring war on Germany. And so the help from across the Atlantic which Churchill so desperately wanted became a reality.

As was the case in the United Kingdom, collection of scrap for recycling was a significant part of the U.S. war effort. School children collected paper, scrap metal, rubber, and rags as part of a campaign claiming to be "the greatest of scrap drives," a campaign which had as its slogan "Get in the Scrap!"[28] The Library of Congress contributed 150 tons of wastepaper and metal.[29] International Harvester instructed its dealers to have their truck drivers who delivered farm machinery to farmers come back with a load of scrap metal, rather than return with an empty truck.[30] The Brooklyn Dodgers had a special game at Ebbets Field in 1942 to which they offered free admission to anyone who brought ten pounds of scrap metal.[31] Jazz great Fats Waller, who wrote "Ain't Misbehaving," encouraged "all us cats to surely do our bit."[32] Even Gypsy Rose Lee, the preeminent burlesque stripper of the era, got involved. At a night club in Detroit, patrons paid an admission fee of one old pot to watch her strip.[33] She wore an unusual costume that night—a rather noisy costume comprised entirely of clanking pots and pans, which she took off one by one. Two strategically placed dippers were the last to go.[34]

In short, collecting materials to be recycled was an endeavor in which just about everyone participated. Most Americans agreed with Institute of Scrap Iron and Steel President Benjamin Schwartz's observation that scrap collecting was a vital, patriotic act that all Americans should do.[35] Recycling had become a civic virtue, not just something that was done for economic reasons.

A deluge of motivational advertisements and slogans encouraged scrap collection for recycling. In one such advertisement, actress Helen Hayes stated, "The kitchen fats make glycerin and glycerin makes the powder charge that drives millions of shells."[36] A widely-distributed pamphlet stated that "an old bucket will make three bayonets" and "the aluminum in a washing machine will build 21 four-pound incendiary bombs." A similar list in a magazine asserted, "One flatiron equals 2 steel helmets or 30 hand grenades . . . . One lawnmower equals 6 three-inch shells. One set of golf clubs equals one .30 caliber machine gun."[37]

There is some debate as to how much scrap collection actually aided the war effort. In some cases, materials that had been collected sat unused, either because of deficiencies in the transportation system or because what was collected exceeded the need for that particular type of scrap.[38] There were other cases, however, in which scrap drives collected critically-needed materials in short supply, thereby contributing significantly to war-time industrial production. For example, a 1942 scrap drive collected 600,000 pounds. of much-needed refined copper.[39] Most important of all, scrap drives contributed to civilian morale by enabling everyone to participate in the war effort. In that regard, they were resounding successes.

A TEMPORARY VIRTUE

The wartime scrap drives in the United States, the United Kingdom, and other countries are superb examples of civic virtue in action. However, this was, as one historian put it, "temporary virtue."[40] Strasser observes with respect to the American experience:

> By the end of the war, Americans were fully ready to favor consumerism over reuse. Rationing, shortages, scrap drives, and home-front propaganda had grown stale. The language of sacrifice and the community participation that initially sustained the campaigns could scarcely compensate for more than fifteen years of thwarted consumption, constrained by depression and war. Scrap drives left many American households without spare pots and pans or aging baby carriages to pass on to young couples. And people had money to spend. In the new peacetime economy, they were ready to cash in their war bonds and satisfy what economists called "pent-up consumer demand."[41]

The story was much the same in the United Kingdom[42] and elsewhere in Europe, once recovery had occurred in these war-torn countries. In short, the greater affluence that many experienced on both sides of the Atlantic in the post-war years gave rise to a throwaway society in which scant attention was paid to recycling.

NOTES

1. "Arch of Constantine" at http://penelope.uchicago.edu/~grout/encyclopaedia_romana/romanurbs/archconstantine.html (last accessed June 13, 2019). In an article entitled "The Arch of Constantine: Propaganda and Ideology in Late Roman Art," Philip Peirce argues that by using reliefs and statues "originally designed to advertise the victories and virtues of Trajan, Hadrian and Marcus Aurelius" Constantine "crudely appropriated the celebration of their deeds as the celebration of his own . . . . The re-cutting of a few busts suggests the deep-rooted continuity of the good ruler throughout the Imperial period" (*Art History*, December 1989, 12:389).
2. "Rome" in *Encyclopaedia Britannica Academic* at http://academic.eb.com/EBchecked/topic/508807/Rome/23920/Slow-decline-of-the-late-empire (last accessed June 13, 2019).
3. The papal states were territories on the Italian peninsula ruled by the pope. At one time, they included a significant portion of present-day Italy. Today, the pope is the temporal ruler only of Vatican City, a 110 acre enclave surrounded by the city of Rome.
4. Paul IV, https://www.ccjr.us/dialogika-resources/primary-texts-from-the-history-of-the-relationship/paul-iv (last accessed June 13, 2019).
5. For additional information about the museum, see http://lnx.museoebraico.roma.it/w/?lang=en (last accessed June 13, 2019).
6. Carole Collier Frick, "The Florentine Rigattieri: Second Hand Clothing Dealers and the Circulation of Goods in the Renaissance," in Alexandra Palmer and Hazel Clark (eds.), *Old Clothes, New Looks: Second Hand Fashion* (New York: Oxford International Publishers Ltd., 2005), 13–17. Florentines made a social distinction between *rigattieri* and *rivenditori*, the latter being old clothes or rag dealers.
7. Andrew Brooks, *Clothing Poverty: The Hidden World of Fast Fashion and Second-Hand Clothes* (London: Zed Books, 2015), 75.
8. Beverly Lemire, "Shifting Currency: The Culture and Economy of the Second Hand Trade in England, c. 1600–1850," in Palmer and Clark (eds.), *Old Clothes, New Looks*, 30.
9. Ibid., 35.
10. Susan Strasser, *Waste and Want: A Social History of Trash* (New York: Henry Holt and Company, 1999), 43.
11. Henry Wadsworth Longfellow wrote "Paul Revere's Ride" in 1860—eighty-five years after the event in question took place. The poem, which begins, "Listen my children and you shall hear, of the midnight ride of Paul Revere . . . ." makes no mention of William Dawes, who also rode from Boston to Lexington, Massachusetts, the evening of April 18, 1775, to warn Samuel Adams and John Hancock that British troops were coming to arrest them ("The Paul

Revere House: The Real Story of Revere's Ride" at https://www.paulreverehouse.org/the-real-story/ [last accessed June 13, 2019]). But then, it would not have been very good poetry if Wadsworth had written, "Listen my children and you shall hear, of the midnight ride of William Dawes . . . ." (The full text of "Paul Revere's Ride" can be accessed at https://poets.org/poem/paul-reveres-ride [last accessed June 13, 2019]).

12. Carl A. Zimring, *Cash for Your Trash: Scrap Recycling in America* (New Brunswick, NJ: Rutgers University Press, 2005), 1.

13. Ibid., 12–36.

14. Ibid., 16.

15. Ibid., 19.

16. Dodd is believed to be the inspiration for one of the characters in Charles Dickens's novel, *Our Mutual Friend* ("Charles Dickens and Henry Dodd" (http://www.charlesdickensinfo.com/novels/our-mutual-friend/charles-dickens-and-henry-dodd/ [last accessed June 13, 2019]).

17. Tim Cooper, "Challenging the 'Refuse Revolution': War, Waste and the Rediscovery of Recycling 1900–50," *Historical Research* (November 2008). 81: 712–715.

18. See, for example, Thomas Carlyle, *The French Revolution: A History*, rev. ed. (New York: The Colonial Press, 1899), 12–13, 264, and "Confederate Cannons, Yankee Church Bells" on the blog of Gettysburg National Military Park at https://npsgnmp.wordpress.com/2012/01/13/confederate-cannons-yankee-church-bells/ (last accessed June 13, 2019). As noted on the Gettysburg National Military Park blog, in April of 1862, Confederate General P.G.T. Beauregard issued an appeal to those living in states that were part of the Confederacy asking them to donate metals to meet the wartime emergency, including bells for the casting of cannon. In the patriotic fervor of the moment, many bells were donated, particularly if they had been cast in the North prior to the war. One of the cannon made from bells that were melted down can be seen at the Gettysburg National Military Park. Cast in Memphis in 1862 by the firm of Quimby and Robinson, this cannon, which shot a twelve-pound cannon ball, is located just to the north of the North Carolina Monument on Seminary Ridge, which was the departure point for Pickett's Charge.

19. A celebrated example of this occurred in New York City during the American War for Independence. In 1770, the grateful residents of New York, celebrating the repeal of the Townshend Acts which had imposed numerous taxes on the American colonists, erected a statue of King George III. Six years later, they decided that they did not like King George III all that much after all. After General George Washington had the U.S. Declaration of Independence read to troops of the Continental Army on July 9, 1776, New Yorkers pulled down the statue, which was melted down to make bullets for Washington's troops ("Statue of King George III Pulled Down, Melted to Make Bullets for Patriots" at http://www.nyfreedom.com/bowlinggreen.htm [last accessed June 13, 2019]).

20. Cooper, "Challenging the 'Refuse Revolution': War, Waste and the Rediscovery of Recycling 1900–50," 716. Cooper notes that in the United Kingdom, it was not until 1918 that the National Salvage Council was created, which was largely exhortatory in nature.

21. "Timeline of the Great Depression" at http://www.hyperhistory.com/online_n2/connections_n2/great_depression.html (last accessed June 13, 2019). $16 billion in 1929 would be the equivalent of $220 billion in 2015.

22. "The Great Depression Statistics" at http://www.shmoop.com/great-depression/statistics.html (last accessed June 13, 2019). It is sometimes assumed that in the depths of the Great Depression, no one had work. That, of course, was not the case. Even in 1933, the worst year of the depression in the United States, three-quarters of the workforce was still employed. Moreover, even though they scaled back on their purchases fearful of what lay ahead, they continued to buy things. Some companies even claimed that their products would save money for those who purchased them. General Electric, for example, said that their refrigerators could save 20 to 30 percent in monthly food bills and offered a time payment plan to make it easier for customers to purchase them (Strasser, *Waste and Want*, 209)

23. "Merchant Navy Losses" at http://www.worldnavalships.com/merchant_navy_losses.htm (last accessed June 13, 2019).

24. M.G. Kendall, "Losses of U.K. Merchant Ships in World War II," *Economica* (1948), 15:289.

25. Quoted by Gary Sheffield in "The Battle of the Atlantic: The U-Boat Peril" at http://www.bbc.co.uk/history/worldwars/wwtwo/battle_atlantic_01.shtml (last accessed June 13, 2019).

26. Cooper, "Challenging the 'Refuse Revolution': War, Waste and the Rediscovery of Recycling 1900–50," 717–718.

27. Quoted in ibid., 718.

28. Strasser, *Waste and Want*, 229–230.

29. Ibid., 240.

30. Ibid., 247.

31. Zimring, *Cash for Your Trash*, 81.

32. Quoted by Zimring in *Cash for Your Trash*, 82.

33. Strasser, *Waste and Want*, 235.

34. Noralee Frankel, *Stripping Gypsy: The Life of Gypsy Rose Lee* (Oxford University Press, 2009), 136. History does not tell us whether men asked their wives if they could take a pot or pan from the kitchen so that they could support the war effort by attending the performance.

35. Zimring, *Cash for Your Trash*, 82.

36. Strasser, *Waste and Want*, 253. The advertisement was actually somewhat misleading. Strasser notes, "In fact, glycerine was used in making nitroglycerine, a component not of ordinary gunpowders but of double-base propellant powders. These were not explosives of major importance . . . ."

37. Ibid., 261.

38. See, for example, Zimring, *Cash for Your Trash*, 94, and Strasser, *Waste and Want*, 235.

39. Donald M. Nelson, *Arsenal of Democracy: The Story of American War Production* (New York: Harcourt, Brace and Co., 1946), 174.

40. Zimring, *Cash for Your Trash*, 100–101.

41. Strasser, *Waste and Want*, 262–263,

42. See, for example, Cooper, "Challenging the 'Refuse Revolution': War, Waste and the Rediscovery of Recycling 1900–50," 716–717.

Chapter Seven

A New Era

In time there was to be a rebirth of the civic virtue of recycling. It took growing environmental concerns to change public consciousness and refocus attention on recycling. Thus, recycling came to done—not for architectural reasons, not simply for economic reasons, not as part of a war effort—but rather as a way of lessening the load that humankind places on the ecosystems that sustain life, a cultural shift that marked the beginning of a new era.

THE LAND ETHIC AND RECYCLING

As noted in the Prologue, in *A Sand County Almanac,* Aldo Leopold states that his land ethic "simply enlarges the boundaries of the community to include soils, waters, plants, and animals, or collectively: the land."[1] He continues, "In short, a land ethic changes the role of *Homo sapiens* from conqueror of the land-community to plain member and citizen of it. It implies respect for his fellow-members, and also respect for the community as such."[2] This includes respect for Mother Earth and the natural resources she provides. It includes practicing good stewardship of the natural resources to which we have access, rather than wasting them by being part of the throw-away society. It includes adopting lifestyles that lesson the burdens that we place on the ecosystems which sustain life.

Unlike the throwaway culture, which in many respects is self-centered and reflects the domination over nature school of thought, the land ethic emphasizes good stewardship and responsible use of that with which we have been entrusted—responsibilities which include lessening the burdens we place on the world's resources and the ecosystems that sustain life, burdens that can be lessened by recycling:

- Recycling aluminum cans saves 95 percent of the energy required to make the same amount of aluminum from bauxite ore; a ton of recycled aluminum saves 40 barrels of oil and 20 cubic yards of landfill space.[3] Recycling just one aluminum can saves enough energy to watch television for three hours.[4]
- One ton of recycled newsprint saves 7,000 gallons of water and 4.6 cubic yards of landfill space.[5] Recycling paper generates 95 percent less air pollution than conventional methods of making paper.[6] Recycling a single run of the Sunday *New York Times* would save 75,000 trees.[7]
- Recycling one ton of plastic saves 16.3 barrels of oil and 30 cubic yards of landfill space.[8] The production of plastic accounts for 4 percent of energy consumption in the United States.[9]
- Recycling glass uses 50 percent less energy, generates 20 percent less air pollution, and generates 50 percent less water pollution than conventional methods of making glass.[10] (In the United States today, over 30 percent of the raw material used in glass production comes from recycled glass.[11])
- One ton of recycled steel saves 1.8 barrels of oil and 4 cubic yards of landfill space.[12] A 60–watt light bulb can be run for more than a day on the amount of energy saved by recycling one pound of steel.[13]

In these ways, and in other ways as well, recycling (which includes repurposing and reusing) conserves natural resources and helps protect the environment, thereby contributing to environmental stewardship.[14]

Recycling is also part of environmental justice. If we simply consume in a self-centered manner, giving no thought to the impact that our lifestyles have on the world's resources and what is left for others to use, we are without realizing it perpetrating environmental injustice. In *The Human Right to a Green Future*, Richard P. Hiskes observes, "As a matter of justice, all citizens present and future possess the environmental human rights to clean air, water, and soil."[15] To this it might be added that all people present and future also have the environmental human rights to have access to the mineral resources and other natural resources essential for their well-being. Recycling, by lessening the load that we place on these resources, helps ensure that others have access to that which is essential for their well-being. Every barrel of oil that is saved by conservation and by recycling is a barrel that is potentially available for future generations. The same is true of iron ore, bauxite ore, and a host of other raw materials.

Moreover, in a certain range of cases, recycling can be good for the bottom line. This is not just because, as the examples noted above illustrate, recycling uses less energy than, say, making aluminum from bauxite ore, or because recycling reduces landfill costs. Recycling can be good for the bottom line in other ways as well. It is instructive to look at the experience of the DuPont's buildings division, which manufactures Corian and Zodiaq quartz solid surfaces (widely used for counter tops) and Tyvek weatherization systems. The company discovered that scraps of Corian could be crushed and made into gravel that could be sold for landscaping. They also discovered that leftover pieces of Tyvek could be shredded and made into new Tyvek with no reduction in the quality of the product. In an interview with the *Gunther Report*, Dave Walter, the DuPont executive who spearheaded the zero waste effort, noted that cafeteria waste was a "tough one" at first, but then they discovered that it could be turned into worm bedding and sold to bait stores, fishermen, and gardeners. Waste that cannot be made into anything else is burned as a fuel to produce energy. As a result of this campaign, the buildings division went from sending 81 million pounds of waste to landfills in 2008 to zero in 2012. The most pleasant surprise, however, was that the company now generates revenue from the recycled products that it sells. This is in addition to the money it saves by avoiding landfill costs.[16]

It should be added that companies like DuPont that recycle do not do this simply because it is good for the bottom line. They also do it because they believe that it is the right thing to do. There are many people like Dave Walter at companies both large and small. People who care about the environment. People who take the social responsibilities of business seriously. The good news in all of this is that good ethics and good business can—and often do—go hand-in-hand.

In some cases, recycling presents new economic opportunities for socially-concerned entrepreneurs who recognize the significance of potential markets for products that are made from recycled materials. Take, for example, Perfect Rubber Mulch, a family-owned company based in Chicago whose motto is "Contributing to a Greener Environment One Tire at a Time." As their name and motto suggest, they recycle old tires to make new products for which there is consumer demand. These include rubber paver tiles for horse paddock areas and for landscaping and shredded rubber for horse arenas.[17] As in the case of DuPont, the family that owns Perfect Rubber Mulch believes that good ethics and good business go hand-in-hand. Most important of all, they put these ideals into practice.

LEADING THE FIELD

While significant progress has been made in the United States in recent years, Switzerland, Germany, Austria, and several other countries have far better records when it comes to recycling. A frequently-cited list of countries with the highest recycling rates shows Switzerland with the highest recycling rate (52 percent), followed by Austria (49.7 percent) and Germany (48 percent), with the United States seventh on the list (31.5 percent).[18] Another list, which was compiled by a manufacturer of recycling bins using information provided by the European Environment Agency and the U.S. Environmental Protection Agency, puts Austria at the top of the list with a 63 percent recycling rate, followed by Germany (62 percent), Belgium and Switzerland (50 percent), the United States and Sweden (49 percent), the Netherlands (48 percent), Luxembourg (46 percent), Norway (42 percent), and Denmark (40 percent).[19]

The bottom line in all of this is that Americans, while doing much better with respect to recycling than was the case in previous years, have much to learn from our friends in Europe.

SWITZERLAND: A CASE STUDY

Recycling is a way of life in Switzerland today. Photographer and author Jessica Cartwright, after moving to Zurich, Switzerland's largest city, observed, "If there is one thing I am constantly in awe of when it comes to living in Switzerland, it is the Swiss' commitment to reducing waste and recycling. The Swiss take recycling to a whole new level . . . . they recycle EVERYTHING. I almost feel bad if I don't pick apart my yogurt containers and recycle the aluminum tops and plastic separately."[20] And indeed there are many who do take the tops off their yogurt containers so that they can recycle the aluminum tops and plastic cups separately.

There are recycling walls at supermarkets with separate slots for plastic milk bottles, PET plastic bottles such as soft drink bottles,[21] aluminum cans, clear glass, brown glass, green glass, and other recyclable items. Many cities and towns also have color-coded recycling bins on sidewalks along streets. Some have underground recycling bins with chimney-like tubes rising above the pavement through which recyclable items can be dropped to the underground bins. As in the case of recycling walls in supermarkets, the tubes are labeled with respect to which types of items should be placed in each of them.[22] In some places, there are bins for organic wastes to be composted with a sign saying "*Bitte nur Kompost* (Please, only compost.")[23] There are containers for cardboard in many neighborhoods. Paper is bundled up and collected monthly, the collection schedule posted on-line. Packaging materials for large items such as computers and television sets can be taken back to the store where the items were purchased. There are various recycling centers in Zurich to which other items can be taken—everything from Styrofoam to the mortal remains of household pets.[24] As in Germany and other countries where there is a strong commitment to recycling, those who fail to recycle are often submitted to a good deal of peer pressure.

Recycling is free in most parts of Switzerland. On the other hand, throwing away rubbish costs money. Each rubbish bag is required to have a sticker on it, which costs at least one euro, sometimes more.[25]

THE TRIDEL S.A. POWER PLANT IN LAUSANNE

Recycling bins and walls in supermarkets are only part of the recycling story in Switzerland. A creative approach to the end use part of the recycling process can be found at the Tridel S.A. combined heat and power plant in Lausanne, a picturesque city of 130,000 located on Lake Geneva. The plant is fueled by garbage. The shareholders of Tridel S.A. include the cities in the catchment area (the area the power generating plant serves), which has a total population of 380,000.[26]

At most power generating plants, much of the heat produced by firing the furnaces that turn the water into steam to run the turbines that power the generators is wasted. At dual function power plants, heat that would otherwise be wasted is used to heat water for use in nearby buildings. Unlike electricity, which can be transported great distances over high voltage transmission lines, heat must be used locally. In the case of the Tridel plant, the excess heat that is produced is used to provide hot water and space heating for a major teaching hospital and for apartments providing housing for 18,000 persons. Insulated pipes in a one kilometer (.62 mile) long tunnel carry hot water to a secondary plant at Pierre-de-Plan that provides heating to the hospital and apartments. A special two-stage design of the plant's turbine allows producing less electricity and more heat in the winter when heat demand is higher with more electricity and less heat produced in the summer. The fuel-to-electricity efficiency is no higher than for any other steam turbine power plant—about 35 percent. But when the thermal efficiency for space and water heating is included in the equation, the efficiency of the plant can be as high as 85 percent. The Tridel plant went into operation in 2006.[27]

Because Lausanne is a hilly city and the Tridel plant is located 250 meters (820 feet) above lower portions of the city, local residents did not want an endless stream of garbage-hauling trucks disrupting the tranquility of their city as the trucks chugged up the hill to the plant. To alleviate these concerns, Tridel built a four kilometer (2.5 mile) long standard gauge railway tunnel under the city so that garbage could be compacted and loaded on specially-constructed train cars to be transported from the unloading area at Sebeillon station to the plant. From higher parts of Lausanne and surrounding communities, trucks haul the garbage directly to the plant. Some of the garbage is shipped in by rail from locations several hundred kilometers away, including from Germany. The garbage does need to be partially presorted with recyclable materials such as aluminum cans and non-combustible materials removed.[28]

Anytime garbage is incinerated, the possibility of air pollution resulting from the flue gasses is a concern. At the Tridel plant, the flue gasses go through an electrostatic precipitator, which removes the bulk of the fly ash. The gasses then go through a series of scrubbers that dissolve acid gases and heavy metal salts, including mercury salts. The water used in the scrubbers for the most part comes from the runoff from rooftops and parking lots after a rainfall, water that is stored in cisterns until it is needed. The precipitator and the scrubbers reduce toxic elements in the flue gasses to less than 10 percent of what is allowed by Swiss and EU regulations, with the exception of nitrogen dioxide, which is reduced to roughly 50 percent of what is allowed by Swiss regulations or 25 percent of what is allowed by EU regulations. The water used in the scrubbers is collected in a tank prior to treatment to remove toxic substances that were scrubbed out of the flue gasses. The fly ash from the precipitator is mixed with the sludge from the scrubbers, which is then mixed with limewater and cinders to form a solid mass prior to being sent by rail to a non-hazardous landfill. The volume of this solid mass is about 10 percent of the volume of the compacted rubbish (or 20 percent by weight).[29]

Revenue produced by the Tridel plant comes from the sale of the electricity and heat produced by the plant and from the waste disposal fees paid by municipalities and others who take their refuse to the plant to be used as fuel. The plant was expensive to build, in part because of the cost of the rail tunnel and the tunnel for the heat pipes taking heat to the secondary plant at Pierre-de-Plan. However, the plant's balance sheet has consistently shown a profit, including in its first year of operation.[30]

SWEDEN'S GARBAGE PROBLEM

Sweden has an unusual garbage problem—it is running out of garbage. That takes a little unpacking. What is somewhat unique about Sweden is that approximately half of the country's household waste is burned in incinerators to heat buildings and power the country's 32 waste-to-energy (WTE) plants that produce electricity.[31] Sweden even imports waste from other countries, with 800,000 tons of trash each year coming from the United Kingdom, Norway, Ireland, and Italy, to help fuel these plants.[32]

Sweden's approach might seem like an elegant solution to what is a very serious problem in other countries. And in certain respects it is. However, Sweden's WTE program is not without its challenges. One is dealing with the pollutants that result from burning garbage. Another is dealing with the ashes which remain after the garbage has been burned. This problem, of course, is not unique to Sweden's WTE facilities. As noted, at the waste-fueled Tridel plant in Lausanne, Switzerland, the flue gasses go through an electrostatic precipitator, which removes the bulk of the fly ash (small particles that are swept up in the flue gasses, which if left untreated are toxic). The flue gasses subsequently go through a series of scrubbers that dissolve acid gases and heavy metal salts, including mercury salts. The fly ash from the precipitator is combined with the sludge from the scrubbers, which is then mixed with limewater and cinders to form a solid mass prior to being sent by rail to a non-hazardous landfill. A similar process, with some variation, is followed at Sweden's WTE plants. At the Filborna plant in Helsingborg, for example, bottom ash (the ash left at the bottom of the combustion chamber, which, unlike fly ash, is not toxic) is sorted for metals that can be recycled with the remaining ash used as fill for road construction and other projects. The fly ash, separated from the flue gasses by electrostatic precipitators, is taken to a landfill certified for hazardous materials.[33] It might be added that in view of the fact that a small amount of the byproducts of the incineration process at the Filborna plant (and other plants as well) end up in landfills, Sweden will not be able to achieve its goal of zero waste without further refinement in the way that fly ash is handled (though because the amount of waste that goes to landfills is quite small, Sweden might still be able to achieve 99.99 percent of its goal of zero waste). When compared to the Filborna plant and other Swedish WTE plants, the Tridel plant in Switzerland does not provide a practical alternative since an even greater portion of the ash at the Tridel plant ends up in landfills than at Swedish WTE plants.

Another challenge facing Sweden is that using such a large portion of waste for fuel means that plastic items and other materials that could be separated and recycled end up being burned. Weine Wiqvist, CEO of the Swedish Waste Management and Recycling Association (*Avfall Sverige*), notes that reprocessing plastic and other recyclable items to make new products uses less energy than burning them and then creating new products from scratch. He comments, "We are trying to 'move up the refuse ladder,' as they say, from burning to material recycling . . . ."[34]

These, however, are manageable problems—problems that are more than offset by the environmental advantages of WTE plants, advantages such as reducing the carbon footprint of power generation and providing a use for refuse that might otherwise be discarded.

MIGHT WTE PLANTS WORK IN LARGER COUNTRIES SUCH AS THE UNITED STATES?

Are WTE plants something that is only feasible in smaller countries such as Sweden and Switzerland? Or might they work in larger countries as well? Actually, though receiving very little publicity, there are a number of WTE plants in the United States. The Energy Recovery Council, a national association representing companies, local governments, and others involved in the U.S. waste-to-energy sector reports that in 2017, there were 75 waste-to-energy facilities operating in 21 states.[35] However, this is a decline from the 84 facilities operating in 23 states in 2014,[36] which was a decline from the 86 facilities operating in 24 states in 2010.[37] Though the decline is gradual, the trend line is clearly going down, rather than up, which suggests that WTE plants are not likely to be the wave of the future in the United States as things currently stand. A carbon tax on fossil fuels might change the economic equations since WTE plants are fueled by garbage, rather than fossil fuels. Whether that happens, however, remains to be seen and is probably unlikely in the current political climate.

SINGLE-STREAM RECYCLING

In recent years, single-stream recycling, which involves putting everything to be recycled in the same recycling bins without separating glass, paper, plastic, etc., has gained popularity in the United States. Take, for example, Boulder County in Colorado. The county's website states, "Boulder County recycling bins have evolved for the 21st Century . . . . The new program is called 'single-stream' recycling. It's the future for responsible resource conservation and an important step toward meeting our goal of building a Zero Waste community by 2030." The website adds, "Single-stream recycling makes it almost as easy to use the recycling bin as it is to use the trash can, so for the previously unconverted, there's no excuse for not recycling."[38] And indeed, convenience is the greatest advantage of single-stream recycling—convenience that can significantly increase the amount of refuse put in recycling bins.

In single-stream recycling systems, once refuse to be recycled is picked up by collection trucks, it is taken to a material recovery facility (MRF, an acronym which, when pronounced, rhymes with "surf.") At the MRF, the refuse is sorted into various categories of recyclable materials. In some simple systems, this is done by hand, much as was done in the dustbins of nineteenth-century England. Somewhat more commonly, the sorting process is facilitated by technologically-sophisticated machines that do much of the work. For example, CP Manufacturing, a major manufacturer of refuse sorting equipment, includes in its product line optical sorting equipment for paper, plastic, glass, and scrap metal.[39] Brent Batliner, a manager at the St. Louis-area recycling division of Republic Services, explains how a particular scanner programmed to identify milk jugs, detergent bottles, and other containers made of high-density polyethylene (HDPE) plastic (No. 2 plastic) works: "It's got infrared scanners that, as material passes underneath it [on the conveyer belt], will read the chemical makeup of the bottle"[40] In this particular system, when a milk jug or detergent bottle reaches the scanner, a row of air jets blows it off the conveyer belt and into a storage bin for HDPE plastic.[41]

Single-stream recycling is not without its limitations. Glass is a particular problem. The Container Recycling Institute, a nonprofit research and advocacy group, explains, "In single-stream programs, it is virtually impossible to prevent glass from breaking as it goes to the curb, is dumped in the truck, gets compacted, gets dumped on the tipping floor of the MRF, is repeatedly driven over by forklifts, and is dumped on conveyor belts to be processed by the MRF."[42] All this broken glass reduces the amount of refuse that gets recycled. It also sometimes contaminates other recyclables, such as bales of paper.[43] Because of the problems that glass causes, some municipalities ban glass from refuse to be picked up for recycling.

Susan Collins, executive director of the Container Recycling Institute, believes that what single-stream recycling gains in volume is sacrificed in loss of quality. She observes, "In terms of preserving the quality of materials so that the maximum materials collected can actually be recycled, single-stream is one of the worst options." She notes that about a quarter of the refuse picked up for single-stream recycling ends up going to the dump—as much as 40 percent in the case of glass.[44] What this means in practical terms is that single-stream recycling is not the path to zero waste envisioned by Boulder County and other counties and municipalities.[45]

The situation is complicated by the fact that China, which was processing much of the waste collected in other countries, in 2018 banned the import of most plastics and other materials heading for their recycling processors (which had been handling nearly half of the world's recyclable waste). This move was intended to halt a deluge of contaminated materials that were costly to recycle. Since then, China's imports of plastics to be recycled have plunged by 99 percent with imports of mixed paper dropping by a third.[46]

Malaysia replaced China as the largest importer of plastic for recycling. But now, Malaysia and other countries in Southeast Asia are imposing blanket bans on imported waste.[47] Other countries are likely to follow.

The situation is further complicated by the fact that in many parts of the United States, the companies that do curbside pickup also operate landfills, which are less expensive and, hence, more profitable for these companies than recycling. The result is that a significant portion of the waste left in curbside containers for recycling does not end up being recycled at all but instead is taken to landfills. This means that even though 75 percent of the waste produced in the United States could be recycled, only about 34 percent ends up being recycled.[48]

This is not universally the case throughout the United States. Minneapolis and Saint Paul, for example, work with independent non-profit recycling companies whose mission is to recycle as much as they can. This results in their sending as much as 90 percent of their high quality processed raw materials to manufacturers within the state.[49]

So where does this leave us? In countries such as Switzerland, where recycling is a strong civic virtue, single-stream recycling makes no sense at all. In countries such as the United States, where the civic virtue of recycling is not as highly-developed, the situation is far more complex. While the facts on the ground vary greatly from municipality to municipality and from region to region, it is plausible to argue that when combined with curbside pick-up, single-stream recycling in a certain range of cases does increase the amount of refuse that is recycled, even if a quarter of what is picked up ends up going to a landfill, rather than being recycled. Yet one must ask, might it not be possible to do better than that? Might not more municipalities be able to follow the examples of Minneapolis and Saint Paul and send a higher percentage of their recyclable wastes to recycling plants to be recycled? And might not we learn from our friends in Switzerland and other European countries that presorting pays dividends and by so doing start moving away from single-stream recycling?

SHOULD RECYCLING BE MANDATORY?

The ideal situation, of course, would be for recycling to be done voluntarily as an expression of civic virtue. But when civic virtue is not sufficient to get the job done, should recycling be mandatory with legal penalties imposed on those who fail to recycle? This is a matter fraught with controversy with strongly-held opposing views on almost every aspect of the debate.

Before looking at this controversy in greater detail, it should be noted that there is a difference between fining people if they fail to recycle, on the one hand, and, on the other hand, charging them disposal fees if their failure to recycle results in their refuse being taken to a landfill.

This distinction is frequently overlooked in popular discussions of these issues. It is not unusual for those who have not thought through these matters carefully to lump together under the rubric of "mandatory recycling" both fines imposed on people if they do not recycle and disposal fees charged those who fail to recycle. There is, however, a fundamental difference between the two. The former is a form of punishing people if they do not do what the government wants them to do. The latter is simply expecting individuals to cover the costs of the decisions that they make.

What the Swiss are doing in this regard is relatively uncontroversial. If someone adds to the cost of waste disposal by not sorting their garbage for recycling, resulting in it being taken to a landfill, it is both fair and appropriate to expect them to pay these additional costs, assuming, of course, that what they are asked to pay is an accurate reflection of the cost of taking refuse to a landfill, including the environmental costs. Fining people or otherwise penalizing them if they fail to recycle, however, is a more complicated matter.

When discussing these matters, it is useful to contrast the *social engineering model* and the *individual rights and compelling public interest model* of the relationship between individuals and society as a whole. The social engineering model focuses primarily on the common good. In this approach, individual rights take second place to promoting what is believed to contribute to the common good. In contrast, the individual rights and compelling public interest model starts with affirmations of individual rights and is supportive of intervention only if there are compelling reasons for doing so. These two models, stripped to their essentials, differ primarily with respect to where the burden of proof lies. The social engineering model starts with some sort of understanding of the common good and places the burden of proof on those who would defend individual rights in the face of what is asserted to contribute to the common good. The individual rights and compelling public interest model begins with an affirmation of individual rights and supports intervention in the lives of private citizens only if there are compelling reasons for intervening, an approach that is more libertarian in its orientation than is the social engineering model. There is wisdom to be found in both models. And indeed, there is something to be said for a creative tension between the two models.

If one simply takes the social engineering approach, it is relatively easy to defend mandatory recycling because there is no question but that, at least in a certain range of cases, recycling conserves resources and by so doing serves the public interest. One proponent of mandatory recycling writes, "Though on a cursory look, the mandatory recycling laws and programs may seem to be an intrusion on the personal liberty of a citizen, they are meant for the larger interest of the society, besides keeping the environment free from pollution and the people, hale and healthy."[50] Mandatory recycling laws, it might be added, tend to work best in cities that are politically quite liberal—in the United States, cities such as San Francisco[51] and Seattle,[52] both bastions of liberalism.[53] This is perhaps not surprising since those at the liberal end of the political spectrum are more likely to be accepting of the social engineering model and the regulation and intrusion on individual liberty that it often entails.[54]

In the case of the individual rights and compelling public interest model, the key question is whether the restriction of liberty that mandatory recycling involves serves a public interest sufficiently compelling to justify restricting liberty. It is worth recalling that even during the crisis of World War II, scrap collection in both the United States and the United Kingdom was, for the most part, voluntary, rather than mandatory.[55] If there ever was a case in which mandatory recycling served a compelling public interest, one is hard-pressed to find any situation more compelling than that of the war effort in the United States and the United Kingdom during World War II. Yet scrap collection and recycling remained for the most part voluntary, rather than mandatory, though, to be sure, there were massive public relations campaigns encouraging citizens to collect scrap for recycling.

Even if one assumes that environmental protection is a compelling public interest—and there is a strong case to be made for suggesting that it is[56]—it is not a foregone conclusion that mandatory recycling is the best way to accomplish this objective. There is often some degree of "pushback" when government tries to force those who fall under its authority to do something they do not wish to do. Mandatory recycling might even, in a certain number of cases, be counterproductive. One analyst who has looked at both the pros and cons of recycling notes, "Anything that is mandatory is going to create rebellion within a community. It is not uncommon for households to just throw away all of their [recyclable] materials because it's easier for them to do it."[57]

Moreover, even under the best of circumstances, mandatory recycling is often a hard sell. Larry West, a writer who covers environmental issues, observes, "Mandatory recycling is a hard sell in the United States, where the economy runs largely along free market lines and landfilling waste remains inexpensive and efficient."[58] Some go on to argue that mandatory recycling imposes costs on consumers.[59] One writer states, "Plain and simple, recycling costs more than landfilling in most locales." This writer adds, however, that many cities have found ways to reduce recycling costs. "They have cut costs by scaling back the frequency of curbside pickups and automating sorting and processing. They've also found larger, more lucrative markets for the recyclables . . . ."[60]

When all things are considered, there is reason to view mandatory recycling with a considerable degree of skepticism. While it is completely reasonable to expect those who do not recycle to cover their fair share of landfill costs and other costs related to their failure to recycle (which in the case of businesses often happens automatically since most businesses pay landfill costs on a per load basis), it is open to question as to whether it is wise to mandate recycling. What might at first glance seem to be a good idea might not end up being such a good idea after all when examined in greater detail. The key question in all of this is how best to foster and nurture recycling as a civic virtue and, by so doing, increase the amount of refuse that is recycled. In the final analysis, mandatory recycling might not be the best way to go about attempting to accomplish this.

THE OTHER PART OF THE EQUATION

Recycling is only part of the equation. When all things are considered, consumption patterns are as important, if not more important, than recycling when it comes to justice between generations, conservation of vital resources, and protecting the ecosystems which sustain life. Very simply stated, if we consume less, our carbon footprints will be reduced in size, regardless of whether or not we recycle. This is, of course, not an invitation to abandon recycling and revert to being a throwaway society. It does serve to remind us, however, that recycling is only part of the picture.

Writing in the first part of the twentieth century, Max Weber (1864–1920) makes reference to Richard Baxter, a Puritan poet and theologian who said that material goods should only lie on the shoulders of a "saint like a light cloak, which can be thrown aside at any moment." Weber, however, goes on to lament, "But fate decreed that the cloak should become an iron cage." Weber is not optimistic about the prospects for escaping from this iron cage.[61]

The explosion in consumer goods the last century-and-a half is mind boggling. In 1876, Alexander Graham Bell patented the telephone, which went into production the following year.[62] In 1885 in Germany, Gottlieb Daimler and Carl Benz attached internal combustion engines they had designed to wheeled vehicles to create the first motorcycle and the first motorcar.[63] In the early part of the twentieth century, Henry Ford, utilizing assembly line mass production techniques, made the automobile available to the growing middle class in the form of the iconic Model T Ford, more than 15 million of which he sold in the United States with nearly 1 million more in Canada and another 250,000 in Great Britain.[64] In 1888, George Eastman brought to the market a simple box camera, which he called the "Kodak" (a name he coined). Twelve years later, he introduced the first "Brownie" camera, which sold for one dollar and was designed to be used by children—a camera that enjoyed considerable popularity.[65] Thomas Alva Edison's long list of inventions during a career spanning more than half a century includes a practical incandescent light bulb, the phonograph, and a motion picture camera.[66]

The list of new consumer goods goes on and on—washing machines, refrigerators, deep freezes, electric stoves, radios, television sets, personal computers, cell phones, and much more. Some of these new products—radios, television sets, personal computers and cell phones, in particular—provided advertisers with new ways of getting their sales pitches out to potential customers. Cleverly designed ads suggest that the new products will bring happiness and greater quality of life. Manufacturers cash in on this never-ending desire for more by bringing to the market new versions of cell phones, laptop computers, digital cameras, and whatever else might be of interest to potential customers, notwithstanding the fact that the older versions are often perfectly capable of doing the tasks that needed to be done. In short, multinational corporations, preoccupied with extracting as much money as possible from us, have tricked us into thinking that we need the newest version of everything that has been invented and that quality of life is determined by the material possessions that we have—a delusion which is fundamentally false.

By allowing ourselves to be victimized by this feeding frenzy, we all, to varying degrees, have fallen victim to what Paul VI in *Populorum Progressio* called "soul-stifling materialism."[67] And by succumbing to this never-ending desire for more, we find ourselves in the iron cage of which Weber spoke, trapped not only by illusions about what constitutes quality of life but also by massive bills run up on credit cards.

Are we condemned to be incarcerated in this iron cage of soul-stifling materialism? That depends on the value choices we make and how we live our lives.

In *Walden,* Henry David Thoreau stated, "Most of the luxuries, and many of the so-called comforts of life, are not only not indispensable, but positive hindrances to the elevation of mankind."[68] However, replicating the austere lifestyle of Thoreau during his Walden years would be difficult, if not impossible. Even if it were possible to replicate Thoreau's Walden experiences, few today—if any—would willingly embrace a lifestyle without indoor plumbing, washing machines, refrigerators, and many other conveniences that are deeply ingrained in our lifestyles. Nor should anyone be expected to do so. In Roman Catholic theology, there is a long-standing distinction between supererogatory acts, on the one hand, and, on the other hand, duties that are incumbent on everyone. Supererogatory acts are morally praiseworthy acts that go beyond the call of duty. In short, they are not duties expected of everyone. Thoreau's austere lifestyle at Walden was morally praiseworthy. At the same time, it should not be viewed as an ideal that everyone should strive to realize in her or his life. Thoreau himself, it should be noted, went to Walden to learn what he could there and then, after two years, two months and two days, returned to a less austere (albeit still modest) lifestyle.

There is a far-ranging continuum between austere asceticism, on one end of the spectrum, and soul-stifling materialism, on the other end of the spectrum. The gift of simplicity is to be found somewhere in the middle of the spectrum, reminiscent of the Golden Mean to which Aristotle made reference.

Of the four traditional Aristotelian virtues (prudence, justice, temperance, and fortitude), temperance is perhaps the greatest of them. Temperance is not abstinence. It does not suggest, for example, that one should never eat ice cream, chocolates, apple pie, or anything else that might tantalize one's taste buds. Rather, the virtue of temperance suggests that we should exercise moderation in the consumption of these joys of life, rather than indulge to the point of excess.

In short, temperance and the notion of moderation that accompanies it do not suggest that we should not acquire any of the consumer goods that are so readily available today. I would be the first to admit that I greatly enjoy the camera equipment I use to take photographs of wildlife and wildflowers. The key, however, is keeping everything in appropriate perspective—or, if you will, keeping everything in balance. Though I have recently added a smaller point-and-shoot camera, which is easier to haul around, my main camera is several years old, which means that there are now newer versions available—newer versions with features my camera does not have. The camera I have, however, is perfectly adequate for the photography that I do. In fact, it has all sorts of features I have never learned how to use, so replacing it with a camera with even more features that I do not need is not something that makes sense.

The same is true of smart phones and other high-tech products. The companies that produce these products and the retailers who sell them, of course, would be very pleased if we were to replace our older versions with new versions. Desire for profits often sweeps all else before it. Succumbing to these pressures, however, is precisely what can land one in the iron cage, as well as contribute to contamination of ecosystems if electronic devices we no longer want are not recycled appropriately.

If we are to avoid being entrapped in the iron cage, there are two key considerations that come into play—sufficiency and affordability. Is that new cellphone, camera, or other device really something that we need? Or are the devices that we already have sufficient to get the job done? And is that new cellphone, camera, or other device something that fits into our budget? Here the virtue of prudence comes into play. Failure to exercise good judgment on Black Friday (the day after Thanksgiving in the United States, which has become an orgy of excess as millions flock to malls and other retail outlets) and on other occasions can land one in the iron cage before he or she realizes what is happening.

In short, asceticism, while in many respects morally praiseworthy, is not the key to dealing with climate change, conservation of natural resources, and the other environmental issues facing us today. The key is moderation in our consumption, which can be realized when the gift of simplicity is discovered, that is to say, the gift of realizing that moderation, rather than extravagance and excess, is what contributes to quality of life, both for us and, to use the words of Francis, for those with whom we share a common home.

There is a certain sense in which less is more. At first glance, that might not appear to make sense, but it does. When we refrain from excessive consumption, we reduce the pressures that our lifestyles place on vital resources, thereby leaving more for other people, both our near neighbors and our distant neighbors, as well as those not yet born. And we help preserve the ecosystems that are essential for our lives and for the lives of other people and other species.

There is more to the equation as well. If we take the earbuds out of our ears and tear ourselves away from our computer screens, we will rediscover the fascinating world of nature and be enriched by the experience. We will rediscover what it means to live in community and share the wonderful gift of life with our family, our friends, and the others whose lives intersect with ours. These are riches that cannot be measured in financial terms. Riches that we have been missing out on. By rediscovering these riches, we will escape from the iron cage to which Weber made reference and enjoy greater measures of joy and happiness and, yes, better quality of life.

SUMMING IT ALL UP

Recycling has a long and complex history, with many different strands winding their way through the centuries. In many cases, such as wartime recycling, collecting scrap to recycle was a temporary virtue, with the throwaway society becoming the order of the day once the crisis had passed. But in time there was to be a rebirth of recycling—not simply for economic reasons, not as part of a war effort—but rather as a way of lessening the load that humankind places on the ecosystems that sustain life, a shift that marked the beginning of a new era. We live in this new era. Both we as individuals and the societies in which we live are better off because of this cultural transformation.

Much has been accomplished. This, however, is not cause for complacency. As in the case of the sustainable agriculture movement and protection of endangered species, there is much that remains to be done. Even in countries such as Sweden, where there is a strong commitment to recycling, the civic virtue of recycling has not yet been realized fully. And so, even as we celebrate the progress that has been made, there are new goals to be attained, new challenges to be met. The successes of the past quarter of a century give reason to believe that these challenges can be met. The seriousness of the environmental problems which face us means that they must be met if humankind and all other forms of life are not to suffer grievous consequences.

Moreover, it is also essential to recognize that recycling is only part of the equation. Our consumption habits are also part of the equation. Exercising moderation in our consumption habits will not only reduce the need for recycling but will also help conserve vital resources and reduce our carbon footprint at a time when climate change has become a matter of great urgency.

NOTES

1. Aldo Leopold, *A Sand County Almanac with Essays on Conservation from Round River* (New York: Ballantine Books, 1966), 239.
2. Ibid., 240.
3. Santa Barbara County Resource Recovery & Waste Management Division, "Why Recycle?" at http://www.lessismore.org/materials/28–why-recycle (last accessed June 14, 2019).

4. United States Environmental Protection Agency (EPA), "Recycling, Energy Conservation, and Community Beautification" at http://www.epa.gov/region03/beyondtranslation/2013BTF/SessionB-Beautification/MichelleFeldman.pdf (last accessed September 1, 2015).

5. Santa Barbara County, "Why Recycle?"

6. U.S. Environmental Protection Agency, "Recycling, Energy Conservation, and Community Beautification"

7. "A Recycling Revolution" at http://www.recycling-revolution.com/recycling-facts.html (last accessed June 14, 2019).

8. Santa Barbara County, "Why Recycle?"

9. U.S. Environmental Protection Agency, "Recycling, Energy Conservation, and Community Beautification"

10. Ibid.

11. Santa Barbara County, "Why Recycle?"

12. Ibid.

13. "A Recycling Revolution."

14. A study funded by the city of Aarhus, Denmark, bears this out. In an article entitled "Assessing Recycling Versus Incineration of Key Materials in Municipal waste: The Importance of Efficient Energy Recovery and Transport Distances" that appeared in *Waste Management* (May 2012), Hanna Merrild, Anna W. Larsen, and Thomas H. Christensen report that "there are environmental benefits when recycling paper, glass, steel and aluminum instead of incinerating it." In their study, the environmental benefits of recycling cardboard and plastic were less clear. Transport distances were a major factor in some cases. They noted, "In many cases, recyclables can be transported fairly long distances without compromising the benefit of recycling, as long as the right means of transport is used . . . . For other materials, means of transport and transport distances should be taken into consideration when assessing the benefits of recycling" (1009–1018).

15. Richard P. Hiskes, *The Human Right to a Green Future: Environmental Rights and Intergenerational Justice* (Cambridge, U.K.: Cambridge University Press, 2009), 2.

16. Marc Gunther, "How DuPont Is Getting to Zero Waste," *Gunther Report*, April 12, 2012, at http://www.greenbiz.com/blog/2012/04/12/how-dupont-getting-zero-waste (last accessed June 14, 2019). DuPont is not the only company that has achieved zero waste-to-landfill at one or more of its facilities. Other examples include RB (whose brand names include Calgon, Woolite, and Clearsil), Procter & Gamble, Nestle USA, Sunny Delight Beverages, Sunpower (a manufacturer of solar panels), Subaru, West Liberty Foods, Unilever North America, Sustainable Waste Solutions, Superior Essex (which operates a copper data cable manufacturing facility), Miller Coors, Mars Chocolate North America, Cargill, Bridgestone Americas, American Packaging, Diamond Packaging, Eaton, J&J Flooring, Geneco (a British recycling and renewable energy company), and Ford Motor Company (Allan Gerlot, "20 Companies with Zero Waste to Landfill Operations," *Waste360*, August 21, 2015, at http://waste360.com/operations/20–companies-zero-waste-landfill-operations#slide-19–field_images-209791 [last accessed September 15, 2015]). See also Sam Donadio, "Eight Companiees That Are Actually Achieving 'Zero Waste,'" *Power Recycling*, September 6, 2017, at https://thinkpowerrecycling.com/8–companies-achieving-zero-waste/ (last accessed June 17, 2019). Zero landfill, it should be added, is not quite the same as recycling. For example, if waste is simply incinerated, that might result in zero landfill, though incineration just to get rid of waste is not a form of recycling (in contrast to using waste to fuel power generating plants). And burning waste just to get rid of it can cause other environmental problems, among them increased air pollution. Even in cases such as Sweden's waste-to-energy plants which burn waste to fuel power generating plants (discussed later in this chapter), incineration is not always the best form of recycling because it might involve burning plastic bottles and other materials that could be recycled.

17. Perfect Rubber Mulch at https://perfectrubbermulch.com/ (last accessed June 14, 2019).

18. "Countries with the Highest Recycling Rates" at http://www.aneki.com/recycling_countries.html (last accessed June 14, 2019). The list is based on data compiled by BBC.

19. Chiara Francavilla, "Britain Outside Recycling Top Ten," *Materials Recycling World*, August 19, 2013, at http://www.mrw.co.uk/news/britain-outside-recycling-top-ten/8652105.article (last accessed January 16, 2019). See also European Environment Agency, "Highest

Recycling Rates in Austria and Germany—But UK and Ireland Show Fastest Increase," March 19, 2013, at http://www.eea.europa.eu/media/newsreleases/highest-recycling-rates-in-austria (last accessed June 14, 2019).

20. Jessica Cartwright, "Recycling in Switzerland" at http://www.expatica.com/ch/insider-views/Recycling-in-Switzerland_102695.html (last accessed June 14, 2019).

21. PET bottles are plastic bottles made of polyethylene terephthalate (PET), which are commonly used for soft drinks, bottled water, and bottled juices and are usually identified by recycling code "1" on the bottom of the bottle ("Facts on PET" at http://www.factsonpet.com/ [last accessed January 16, 2019]).

22. Mamiko, "A Primer on Recycling in Switzerland," January 28, 2012, at http://www.newlyswissed.com/a-primer-on-recycling-in-switzerland-2/ (last accessed June 14, 2019).

23. Cartwright, "Recycling in Switzerland."

24. Ibid.

25. Imogene Foulkes, "Switzerland" in "Recycling Around the World."

26. Brian Ellis, "Not Wasting Waste," *Engineering & Technology* (June 4, 2008) at http://eandt.theiet.org/magazine/2008/07/not-wasting-waste.cfm (last accessed January 16, 2019). See also "Enhancement of the Value of Waste: A Visit to Tridel SA, Lausanne," *World Environment and Energy* (2011) at http://www.cypenv.info/worldee/files/waste/tridel.aspx (last accessed January 16, 2019).

27. Ellis, "Not Wasting Waste."

28. Ibid.

29. "Enhancement of the Value of Waste: A Visit to Tridel SA, Lausanne."

30. Ibid. For additional information about the Tridel plant, see "Lausanne/Switzerland: Energy-from-Waste Plant" at http://www.hz-inova.com/cms/wp-content/uploads/2014/11/hzi_ref_lausanne-en.pdf (last accessed June 14, 2019).

31. Jonas Freden, "The Swedish Recycling Revolution" at https://sweden.se/nature/the-swedish-recycling-revolution/ (last accessed June 14, 2019).

32. Alanna Ketler, "Sweden Runs Out of Garbage," November 26, 2014, at http://www.collective-evolution.com/2014/11/26/sweden-runs-out-of-garbage-only-1–ends-up-in-landfills/ (last accessed June 14, 2019).

33. Dan Haugen, "Is Burning Garbage Green? In Sweden, There's Little Debate," *Midwestern Energy News*, October 17, 2013, at http://midwestenergynews.com/2013/10/17/is-burning-garbage-green-in-sweden-theres-little-debate/ (last accessed June 14, 2019). In order to encourage recycling, Helsingborg has fitted public waste bins with sound systems that play pleasant music in order to encourage people to recycle (Freden, "The Swedish Recycling Revolution").

34. Quoted by Freden in "The Swedish Recycling Revolution."

35. *The 2017 ERC Directory of Waste-to-Energy Facilities*, 3, at http://energyrecoverycouncil.org/wp-content/uploads/2019/01/ERC-2018–directory.pdf (last accessed June 14, 2019).

36. *The 2014 ERC Directory of Waste-to-Energy Facilities*, 3, at http://energyrecoverycouncil.org/wp-content/uploads/2016/01/ERC_2014_Directory.pdf (last accessed February 23, 2017).

37. *The 2010 ERC Directory of Waste-to-Energy Facilities* at https://www.researchgate.net/publication/265561751_The_2010_ERC_Directory_of_Waste-to-Energy_Plants (last accessed December 18, 2010).

38. Boulder County, "Single-Stream Recycling in Boulder County" at http://www.ecocycle.org/recycle-compost-reuse/singlestream (last accessed June 14, 2019).

39. CP Manufacturing, "Optical Sorting Equipment" at http://www.cpmfg.com/recycling-equipment/optical-sorting-equipment/ (last accessed June 14, 2019).

40. Quoted by Veronique LaCapra in "With 'Single-Stream' Recycling, Convenience Comes at a Cost," *National Public Radio (NPR)*, March 31, 2015, at http://www.npr.org/2015/03/31/396319000/with-single-stream-recycling-convenience-comes-at-a-cost (last accessed June 14, 2019).

41. Ibid.

42. Quoted by Sarah Laskow in "Single-Stream Recycling Is Easier for Consumers, but Is It Better?" *The Atlantic*, September 18, 2014. Additional information about the Container Recycling Institute can be accessed at http://www.container-recycling.org/index.php/home/aboutus.

43. Laskow, "Single-Stream Recycling Is Easier for Consumers, but Is It Better?"

44. Quoted by LaCapra in "With 'Single-Stream' Recycling, Convenience Comes at a Cost." See also Nora Goldstein, "Pros and Cons of Single Stream Recycling," *BioCycle*, July 2003, and Dan Emerson, "Single Stream vs. Source Separated Recycling," *BioCycle*, March 2004.

45. In an editorial in *BioCycle*, Nora Goldstein, who is the journal's editor, notes that many of the companies that have been given contracts for curbside recycling pickup also own landfills. She argues, "Essentially in this business model, the heavy hitters earn high profits on a per ton basis by controlling the waste from the curb. And the reality is that at the end of the day, the best ROI (return on investment) comes from getting waste into the landfill, which is the least expensive and most profitable option for these companies" (Nora Goldstein, "Calling a Spade a Spade," *BioCycle*, June 2015). To the extent that this is the case, there are powerful incentives for waste collection companies with curbside recycling contracts to hold down the amount that is actually recycled and maximize the amount of refuse that goes to landfills that they own, thus thwarting any effort to make zero waste a reality.

46. Cheryl Katz, "Piling Up: How China's Ban on Importing Waste Has Stalled Global Recycling," *Yale Environment 360*, March 7, 2019, at https://e360.yale.edu/features/piling-up-how-chinas-ban-on-importing-waste-has-stalled-global-recycling (last accessed June 14, 2019).

47. Mike Ives, "Recyclers Cringe as Southeast Asia Says It's Sick of the West's Trash," *New York Times.*

48. Erin Biba, "Everything Americans Think They Know about Recycling Is Probably Wrong," *NBC News*, April 14, 2019, at https://www.nbcnews.com/think/opinion/everything-americans-think-they-know-about-recycling-probably-wrong-ncna994261 (last accessed June 14, 2019).

49. Ibid.

50. Rama Lingam, "Mandatory Recycling Laws: Pros and Cons" at https://knoji.com/mandatory-recycling-laws-pros-and-cons (last accessed June 14, 2019).

51. For a narrative of the San Francisco policy, see City of San Francisco Department of Public Works, "Refuse Collection and Disposal Rates (Garbage Rates" at https://sfpublicworks.org/refuserates (last accessed June 14, 2019).

52. City of Seattle Public Utilities, "Ordinance http://clerk.seattle.gov/~archives/Ordinances/Ord_121372.pdf (last accessed June 14, 2019).

53. *Forbes* lists Seattle as the third most liberal city in the United States and San Francisco as the most liberal city (Naomi Shavin, "The Most Conservative and Most Liberal Cities in America," *Forbes* at http://www3.forbes.com/leadership/the-most-conservative-and-most-liberal-cities-in-america/21/ [last accessed January 16, 2019]). See also, Anne Stanley, "Sustainable America: West Coast Cities Leading the Way with Recycling," *American City & County*, March 4, 2015, at http://web.a.ebscohost.com/ehost/detail/detail?sid=cdf5ba5a-d32d-450a-9133–b68495afb908%40sessionmgr4004&vid=11&hid=4207&bdata=JnNpdGU9Z Whvc3QtbGl2ZQ%3d%3d#AN=101387043&db=a9h (last accessed January 16, 2019).

54. Some cities have taken a "meet in the middle" approach with recycling mandatory for businesses but not for individual residents (Michelle Talsma Everson, "Mandate vs. Volunteer: What Works Better for Recycling?" *Earth911*, September 21, 2009, at http://www.earth911.com/inspire/getting-involved/mandate-vs-volunteer-what-works-better-for-recycling/ (last accessed January 16, 2019).

55. In the United Kingdom, the government created a salvage department that was part of the Ministry of Supply to encourage salvaging on the local level during World War II. Initially the Ministry of Supply simply relied on exhortation in an effort to encourage recycling, but when it became apparent that many local authorities were ignoring the call to salvage, the government resorted to compulsory measures, requiring that local authorities participate in salvage efforts (Cooper, "Challenging the 'Refuse Revolution': War, Waste and the Rediscovery of Recycling 1900–50," 712–715). This, however, did not extend to requiring private

citizens to recycle. In the United States, some employers (e.g., International Harvester) required that their employees participate in scrap collection and recycling (Strasser, *Waste and Want*, 247). This was a company initiative, though, rather than something mandated by the government.

56. The savings in the consumption of energy and natural resources noted earlier in this chapter are the building blocks for a fairly compelling case that can be made for recycling.

57. Crystal Lombardo, "Mandatory Recycling Pros and Cons," April 15, 2015, at http://www.visionlaunch.com/mandatory-recycling-pros-and-cons/ (last accessed January 16, 2019).

58. Quoted by Everson in "Mandate vs. Volunteer."

59. See, for example, Grant W. Schaumburg, Jr., and Katherine T. Doyle, *Consumers' Research Magazine*, April 1994, at http://web.a.ebscohost.com/ehost/detail/detail?sid=cdf5ba5a-d32d-450a-9133–b68495afb908%40sessionmgr4004&vid=12&hid=4207&bdata=JnNpdGU9ZWhvc3QtbGl2ZQ%3d%3d#AN=9404187524&db=a9h (last accessed September 14, 2015).

60. "Why is Recycling Not Mandatory in All U.S. Cities?" *Earthtalk* at http://environment.about.com/od/recycling/a/must_recycle.htm (last accessed January 16, 2019).

61. Max Weber, *The Protestant Ethic and the Spirit of Capitalism*, tr. Talcott Parsons (New York: Charles Scribner's Sons, 1958), 181–183.

62. David Hochfelder, "Alexander Graham Bell: American Inventor" at https://www.britannica.com/biography/Alexander-Graham-Bell (last accessed June 14, 2019).

63. "The Industrial Revolution (1750–1900)," *Encyclopædia Britannica* at https://www.britannica.com/technology/history-of-technology/The-Industrial-Revolution-1750–1900 (last accessed June 14, 2019).

64. Carol W. Gelderman, "Henry Ford: American Industrialist," *Encyclopædia Britannica* at https://www.britannica.com/biography/Henry-Ford (last accessed June 14, 2019).

65. "George Eastman: American Inventor, Entrepreneur, and Manufacturer," *EncyclopædiaBritannica* at https://www.britannica.com/biography/George-Eastman (last accessed June 14, 2019)

66. Robert E. Conot and Matthew Josephson, "Thomas Edison: Inventor," *Encyclopædia Britannica* at https://www.britannica.com/biography/Thomas-Edison (last accessed June 14, 2019).

67. Paul VI, *Populorum Progressio*, 18, at http://w2.vatican.va/content/paul-vi/en/encyclicals/documents/hf_p-vi_enc_26031967_populorum.html (last accessed June 14, 2019)

68. Henry David Thoreau, "Economy" in *Walden and Other Writings of Henry David Thoreau,* ed. Brooks Atkinson (New York: The Modern Library, 1950), 13.

Epilogue

In certain respects, the 2016 election in the United States changed everything. In other respects, it changed nothing. That takes a little unpacking. The election placed in positions of power individuals who are motivated far more by love of money than by love of the world of nature. As is apparent to anyone who has watched events in Washington unfold, the Trump administration does not have the same level of commitment to environmental preservation as have preceding administrations dating back to the Nixon era.

What has happened in Washington and elsewhere, however, does not absolve us of the responsibility to be good environmental citizens, a responsibility which includes, but is not limited to, the following:

- *Treating all members of the communities of which we are a part with respect and dignity.* This includes not only our fellow human beings but also all other creatures that are part of the ecosystems which sustain life;
- *Practicing good stewardship with respect to the land and other resources with which we have been entrusted.* This responsibility is incumbent not just on those of us who own property but on others as well. Everyone both can and should practice good stewardship of the earth's resources by recycling, driving fuel-efficient automobiles, opting for modest lifestyles, and otherwise reducing the carbon footprint we leave on planet earth—changes that are essential if the effects of climate change, which are not just something that will happen in the future but are something that is happening right now, are to be mitigated;
- *Supporting organizations and candidates for office who are committed to protecting the environment which is so crucial, not just for humankind, but for wildlife and other members of the biotic communities of which we are a part.* Organizations which play constructive roles with respect to these

> matters include the Audubon Society, the Sierra Club, and Defenders of Wildlife, as well as many other groups and organizations. Their work has assumed even greater significance now that love of money is trumping environmental preservation at the highest levels of government.

With respect to protection of endangered species, the sustainable agriculture movement, recycling, and many other crucial areas as well, cultural transformation is an essential factor that has contributed—and will continue to contribute—to changes that protect and preserve the life-sustaining ecosystems of which we are part. Yes, government does have a role to play—as a facilitator and sometimes as a regulator. But that is not where the story begins or where it ends. If there is to be cultural transformation, including cultural transformation that is essential for dealing with urgent matters such as climate change, it must begin on the grassroots level. It must begin with each of us. This is far more important than anything that happens (or does not happen) in Washington or any other seat of government. In the final analysis, hope is often where we least expect to find it—in each of us. Once we come to this realization, there is reason for optimism about the future.

Bibliography

100 Years of U.S. Consumer Spending: Data for the Nation, New York City, and Boston, U.S. Bureau of Labor Statistics (May 2006), 3, 22.

"Agriculture and Farm Policy," USDA Economic Research Service (June 2005) at http://ageconsearch.umn.edu/bitstream/59390/2/eib3.pdf (last accessed June 11, 2019).

Agriculture Sustainability Institute of the University of California—Davis. "What Is Sustainable Agriculture?" at http://www.sarep.ucdavis.edu/ (last accessed January 17, 2013).

"Arch of Constantine" at http://penelope.uchicago.edu/~grout/encyclopaedia_romana/romanurbs/archconstantine.html (last accessed June 13, 2019).

Associated Press. "U.S. Interior Department Denies Endangered Species Act Protection to Sage Grouse," September 22, 2015, at http://www.nbcnews.com/news/us-news/u-s-interior-department-denies-endangered-species-act-protection-sage-n431701 (last accessed June 10, 2019).

Augustine. *Contra Faustum.* Translated by Richard Stothert. Included in *A Select Library of the Nicene and Post-Nicene Fathers of The Christian Church*, Vol. IV, *Writings in Connection with the Manichaean Controversy.* Edited by Philip Schaff. Grand Rapids, Michigan: Wm. B. Eerdmans Publishing Company, 1887.

Battistoni, Alyssa. "America Spends Less on Food Than Any Other Country." *Mother Jones*, February 1, 2012.

Beacham, Walter, editor. *The Official World Wildlife Fund Guide to Extinct Species of Modern Times*, Volume I. Beacham Publishing Corp., 1997.

Beck, Eckardt C. "The Love Canal Tragedy." *EPA Journal*, January 1979.

Berkeley Youth Alternatives. "Our Mission" at http://www.byaonline.org/about-us/our-mission/our-mission (last accessed April 22, 2019).

Berry, Wendell. "The Clearing" in *Clearing*. New York: Harcourt Brace Jovanovich, 1977.

Berry, Wendell. *Jayber Crow*. New York: Counterpoint, 2000.

Biba, Erin. "Everything Americans Think They Know about Recycling Is Probably Wrong." *NBC News*, April 14, 2019, at https://www.nbcnews.com/think/opinion/everything-americans-think-they-know-about-recycling-probably-wrong-ncna994261 (last accessed June 14, 2019).

"Bison." *West Virginia (WV) Cyclopedia* at http://www.wvexp.com/index.php?title=Bison&oldid=27838 (last accessed June 10, 2019).

Bond, Jack J., and Dwaine E. Umberger, *Technical and Economic Causes of Productivity Changes in U.S. Wheat Production, 1949–1976*, U.S. Department of Agriculture Technical Bulletin 1598 (1979).

"Bone China." *Encyclopaedia Britannica Academic* at http://academic.eb.com./EBchecked/topic/72917/bone-china?sections=72917main&cit=mla&view=print (last accessed September 28, 2015).

Boulder County. "Single-Stream Recycling in Boulder County" at http://www.ecocycle.org/recycle-compost-reuse/singlestream (last accessed June 14, 2019).

"Bradenton Gardens Park" at http://www.bradnergardenspark.org/index.html (last accessed April 22, 2019).

Bradley, Edward. "No Till Farming" at http://www.notillfarming.co.uk/index.html (last accessed June 12, 2019).

Brooks, Andrew. *Clothing Poverty: The Hidden World of Fast Fashion and Second-Hand Clothes*. London, U.K.: Zed Books, 2015.

Burgess, Bonnie B. *Fate of the Wild: The Endangered Species Act and the Future of Biodiversity.* Athens, Georgia: The University of Georgia Press, 2001.

Callicott, J. Baird. *In Defense of the Land Ethic: Essays in Environmental Philosophy.* Albany, New York: State University of New York Press, 1989.

Carlson, Paul H. *The Plains Indians*. College Station, Texas: Texas A&M University Press, 1998.

Cartwright, Jessica. "Recycling in Switzerland" at http://www.expatica.com/ch/insider-views/Recycling-in-Switzerland_102695.html (last accessed June 14, 2019).

Center for Biological Diversity. "Restoring the Gray Wolf" at http://www.biologicaldiversity.org/campaigns/gray_wolves/ (last accessed September 3, 2016).

Centers for Disease Control and Prevention (CDC), "Antibiotic/Antimicrobial Resistance (AR/AMR" at https://www.cdc.gov/drugresistance/ (last accessed June 11, 2019).

Chain Reaction III: How Top Restaurants Rate on Reducing Use of Antibiotics in their Meat Supply, 2017, at https://1bps6437gg8c169i0y1drtgz-wpengine.netdna-ssl.com/wp-content/uploads/2017/09/Chain-Reaction-III.pdf (last accessed June 12, 2019).

Chase, Candace. "Kalispell Founder's Granddaughter Dies." *Daily Interlake* (Kalispell, Montana), August 6, 2010 at http://www.dailyinterlake.com/members/kalispell-founder-s-granddaughter-dies/article_5c14a7e0-a355-11df-bce5-001cc4c002e0.html (last accessed June 10, 2019).

Chicago Park District. "Community Gardens" at https://www.chicagoparkdistrict.com/parks-facilities/community-garden (last accessed April 22, 2019).

Childress, James F. "Just-War Criteria" in *War or Peace? The Search for New Answers*. Edited by Thomas A. Shannon. Maryknoll, New York: Orbis Books, 1980.

Chow, Denise. "One Million Species under Threat of Extinction because of Humans, Biodiversity Report Finds," *NBC News*, 6 May 2019 at https://www.nbcnews.com/mach/science/1-million-species-under-threat-extinction-because-humans-report-finds-ncna1002046 (last accessed June 10, 2019).

City of Minneapolis. "Homegrown Minneapolis Community Gardens" at http://www.minneapolismn.gov/sustainability/homegrown/WCMS1P-129871 (last accessed June 10, 2019).

City of San Francisco Department of Public Works. "Refuse Collection and Disposal Rates (Garbage Rates)" at https://sfpublicworks.org/refuserates (last accessed June 14, 2019).

City of Seattle Public Utilities. "An Ordinance Relating to Seattle Public Utilities" at http://clerk.seattle.gov/~archives/Ordinances/Ord_121372.pdf (last accessed June 14, 2019).

Civil War Trust. "John Singleton Mosby" at http://www.civilwar.org/education/history/biographies/john-singleton-mosby.html?referrer=https://www.google.com/ (last accessed June 10, 2019).

Cohen, Patricia. "$2 Billion Verdict Against Monsanto Is Third to Find Roundup Caused Cancer," *New York Times*, May 13, 2019.

"Compensation Rules for Depredation Incidents" at http://wdfw.wa.gov/conservation/gray_wolf/livestock/compensation.html (last accessed October 21, 2015).

Congressional Acts Pertaining to Yellowstone at http://www.yellowstone-online.com/history/yhfour.html (last accessed June 10, 2019).

Conot, Robert E., and Matthew Josephson. "Thomas Edison: Inventor." *Encyclopædia Britannica* at https://www.britannica.com/biography/Thomas-Edison (last accessed June 14, 2019).

Cooper, Tim. "Challenging the 'Refuse Revolution': War, Waste and the Rediscovery of Recycling 1900–50." *Historical Research* (November 2008), 81: 712–715.

Cornell, Joseph D., and Michelle Miller, "Slash and Burn," *Encyclopedia of Earth*. Edited by Cutler J. Cleveland. Washington, D.C.: National Council for Science and the Environment, 2011.

"Countries with the Highest Recycling Rates" at http://www.aneki.com/recycling_countries.html (last accessed June 14, 2019).

Cutler J. Cleveland. Washington, D.C.: National Council for Science and the Environment, 2011 at http://www.eoearth.org/article/Slash_and_burn (last accessed May 7, 2018).

CP Manufacturing. "Optical Sorting Equipment" at http://www.cpmfg.com/recycling-equipment/optical-sorting-equipment/ (last accessed June 14, 2019).

Cross, Charlene. *The Undying West: A Chronicle of Montana's Camas Prairie*. Golden, Colorado: Fulcrum Publishing, 1999.

Davenport, Carol. "Trump Drilling Plan Threatens 9 Million Acres of Sage Grouse Habitat," *NewYork Times*, December 6, 2018.

Defenders of Wildlife. "Basic Facts about Bison" at http://www.defenders.org/bison/basic-facts (last accessed June 10, 2019).

Defenders of Wildlife. "Mission and Vision" at http://www.defenders.org/mission-and-vision (last accessed June 10, 2019).

Defenders of Wildlife. "Basic Facts about Bald Eagles" at http://www.defenders.org/bald-eagle/basic-facts (last accessed June 10, 2019).

Defenders of Wildlife. "Gray Wolf Recovery Infographic" at http://www.defenders.org/the-war-on-wolves/infographic (last accessed June 10, 2019).

"Defenders Shifts Focus to Wolf Coexistence Partnerships," August 20, 2010, at http://www.defenders.org/press-release/defenders-shifts-focus-wolf-coexistence-partnerships (last accessed June 10, 2019).

Derpsch, Rolf, Theodor Friedrich, Amir Kassam, and Li Hongwen. "Current Status of Adoption of No-Till Farming in the World and Some of Its Main Benefits." *International Journal of Agricultural and Biological Engineering* (March 2010) at http://www.fao.org/ag/ca/CA-Publications/China_IJABE.pdf (last accessed June 12, 2019).

DiCerto, Joseph J. "William F. Cody." *Encyclopaedia Britannica Academic* at http://academic.eb.com./EBchecked/topic/124078/William-F-Cody (last accessed September 28, 2015).

"Dimethoate" at http://pan-uk.org/pestnews/Actives/dimethoa.htm (last accessed February 1, 2013).

Dimitri, Carolyn, Anne Effland, and Neilson Conklin, "The 20th Century Transformation of U.S. Agriculture and Farm Policy," USDA Economic Research Service (June 2005) at http://ageconsearch.umn.edu/bitstream/59390/2/eib3.pdf (last accessed June 11, 2019).

Dodge, Richard Irving. *The Plains of the Great West and Their Inhabitants*. New York: G.P. Putnam's Sons, 1876.

Donadio, Sam. "Eight Companies That Are Actually Achieving 'Zero Waste." *Power Recycling*, September 6, 2017, at https://thinkpowerrecycling.com/8-companies-achieving-zero-waste/ (last accessed June 17, 2019).

"Economic Impact of 2013 BM Sage Grouse Conservation Plan," March 1, 2014, at http://www.westernenergyalliance.org/sites/default/files/Sage%20Grouse%20Economic%20Report%20-%20Final%20from%20Minuteman%20Press.pdf (last accessed June 10, 2019).

Ellis, Brian. "Not Wasting Waste." *Engineering & Technology*, June 4, 2008 at http://eandt.theiet.org/magazine/2008/07/not-wasting-waste.cfm (last accessed January 16, 2019).

Emerson, Dan. "Single Stream vs. Source Separated Recycling." *BioCycle*, March 2004.

Endangered Species Act of 1973 at http://www.fws.gov/endangered/laws-policies/section-3.html (last accessed June 10, 2019).

ERC Directory of Waste-to-Energy Facilities (2010) at https://www.researchgate.net/publication/265561751_The_2010_ERC_Directory_of_Waste-to-Energy_Plants (last accessed December 18, 2010).

ERC Directory of Waste-to-Energy Facilities (2014) at http://energyrecoverycouncil.org/wp-content/uploads/2016/01/ERC_2014_Directory.pdf (last accessed February 23, 2017).

ERC Directory of Waste-to-Energy Facilities (2017) at http://energyrecoverycouncil.org/wp-content/uploads/2019/01/ERC-2018-directory.pdf (last accessed June 14, 2019).

European Commission. "Inspection and Certification—Organic Farming" at http://ec.europa.eu/agriculture/organic/consumer-confidence/inspection-certification_en (last accessed June 12, 2019).

European Environment Agency. "Highest Recycling Rates in Austria and Germany—But UK and Ireland Show Fastest Increase," March, 19, 2013, at http://www.eea.europa.eu/media/newsreleases/highest-recycling-rates-in-austria (last accessed June 14, 2019).

"EU Seeks to Protect Bees from Agricultural Insecticides." *Europe Online Magazine*, January 31, 2013, at http://en.europeonline-magazine.eu/eu-seeks-to-protect-bees-from-agricultural-insecticides_261829.html (last accessed June 12, 2019).

Everson, Michelle Talsma. "Mandate vs. Volunteer: What Works Better for Recycling?" *Earth911*, September 21, 2009, at http://www.earth911.com/inspire/getting-involved/mandate-vs-volunteer-what-works-better-for-recycling/ (last accessed January 16, 2019).

Extension Toxicology Network. "Glyphosate" at http://pmep.cce.cornell.edu/profiles/extoxnet/dienochlor-glyphosate/glyphosate-ext.html (last accessed June 12, 2019).

"FDA To Phase Out Non-Therapeutic Use of Antibiotics in Food Animals," *Medical News Today*, December 12, 2013 at http://www.medicalnewstoday.com/articles/270010.php (last accessed October 23, 2015).

Food and Agriculture Organization (FAO) of the United Nations. "Genetically Modified Crops" at http://www.fao.org/3/i2490e/i2490e04d.pdf (last accessed June 12, 2019).

Food and Agriculture Organization (FAO) of the United Nations. "Sustainable Agriculture and Rural Development" at http://www.fao.org/docrep/U8480E/U8480E0l.htm (last accessed January 15, 2019).

Forbush, Edward Howe. "Extinction of the American Passenger Pigeon" at http://www.wildbirds.org/apidesay.htm. The essay originally appeared in a publication of the Massachusetts Board of Agriculture.

Francavilla, Chiara. "Britain Outside Recycling Top Ten." *Materials Recycling World*, August 19, 2013, at http://www.mrw.co.uk/news/britain-outside-recycling-top-ten/8652105.article (last accessed January 16, 2019).

Francis. *Laudato Si': On Care for Our Common Home*, May 24, 2015, at http://w2.vatican.va/content/francesco/en/encyclicals/documents/papa-francesco_20150524_enciclica-laudato-si.html (last accessed April 30, 2019).

Frankel, Noralee. *Stripping Gypsy: The Life of Gypsy Rose Lee.* New York: Oxford University Press, 2009.

Freden, Jonas. "The Swedish Recycling Revolution" at https://sweden.se/nature/the-swedish-recycling-revolution/ (last accessed June 14, 2019).

Frick, Carole Collier. "The Florentine Rigattieri: Second Hand Clothing Dealers and the Circulation of Goods in the Renaissance," in *Old Clothes, New Looks: Second Hand Fashion*, edited by Alexandra Palmer and Hazel Clark. New York: Oxford International Publishers Ltd., 2005.

Garretson, Martin S. *The American Bison: The Story of Its Extermination as a Wild Species and Its Restoration under Federal Protection.* New York: New York Zoological Society, 1938.

Gasnier, Celine, *et al.*, "Glyphosate-Based Herbicides Are Toxic and Endocrine Disruptors in Human Cell Lines," *Toxicology* (2009), 262: 184–191.

Gelderman, Carol W. "Henry Ford: American Industrialist," *Encyclopædia Britannica* at https://www.britannica.com/biography/Henry-Ford (last accessed June 14, 2019).

"George Eastman: American Inventor, Entrepreneur, and Manufacturer." *EncyclopædiaBritannica* at https://www.britannica.com/biography/George-Eastman (last accessed June 14, 2019).

Goldstein, Nora. "Calling a Spade a Spade." *BioCycle*, June 2015.

Goldstein, Nora. "Pros and Cons of Single Stream Recycling." *BioCycle*, July 2003.
Goold, Bill. "The Madison Buffalo Jump: Rich in Stories and Grandeur." *Distinctly Montana*, September 19, 2015, at http://www.distinctlymontana.com/buffalo154#utm_sguid=89802,7678f8d7-d6f9-b973-5c8e-cc093b54ae3a (last accessed June 10, 2019).
"GPS.Gov: Agricultural Applications" at http://www.gps.gov/applications/agriculture/ (last accessed June 12, 2019).
"The Great Depression Statistics" at http://www.shmoop.com/great-depression/statistics.html (last accessed June 13, 2019).
Grotius Hugo. *The Law of War and Peace (De Jure Belli ac Pacis Libri Tres).* Translated by Francis W. Kelsey. Indianapolis, Indiana: The Bobbs-Merrill Company, Inc., 1925.
Guarino, Ben. "Thousands of Montana Snow Geese Die after Landing in Toxic, Acidic Mine Pit." *Washington Post* (Washington, D.C.), December 7, 2016.
Gunther, Mark. "How DuPont Is Getting to Zero Waste." *Gunther Report*, April 12, 2012, at http://www.greenbiz.com/blog/2012/04/12/how-dupont-getting-zero-waste (last accessed June 14, 2019).
Haberman, Clyde. "For Gray Wolves, a Success Story Not Without Detractors." *New York Times,* November 2, 2014.
Hakim, Danny. "Glyphosate, Top-Selling Weed Killer, Wins E.U. Approval for 5 Years," *New York Times*, November 27, 2017.
Hamilton, Alexander. "The Lack of a Bill of Rights." *Federalist No. 84* in Alexander Hamilton, *Federalist No. 84* in Alexander Hamilton, James Madison, and John Day, *The Federalist.* Edited by Benjamin Fletcher Wright. Cambridge, MA: The Belknap Press of Harvard University Press, 1961.
Harbison, Martha. "What's up with the Greater Sage Grouse? A Definitive Guide to the Most Talked-About Bird in the West," *Audubon Newsletter*, May 29, 2015.
Hardwell, Lennart, and Mikael Erikson, "A Case-Controlled Study of Non-Hodgkin Lymphona and Exposure to Pesticides," *Journal of the American Cancer Society* (March 15, 1999), 85: 1353–1300.
Hart, John Fraser. "Change in the Corn Belt" at http://www.geo.hunter.cuny.edu/courses/geog347/articles/change_belt.pdf (last accessed April 22, 2019).
Haugen, Dan. "Is Burning Garbage Green? In Sweden, There's Little Debate." *Midwestern Energy News*, October 17, 2013, at http://midwestenergynews.com/2013/10/17/is-burning-garbage-green-in-sweden-theres-little-debate/ (last accessed June14, 2019).
Hawthorne, Nathaniel. *Mosses from an Old Manse*. Columbus, Ohio: Ohio State University Press, 1974.
Hiskes, Richard P. *The Human Right to a Green Future: Environmental Rights and Intergenerational Justice.* Cambridge, U.K.: Cambridge University Press, 2009.
Hochfelder, David. "Alexander Graham Bell: American Inventor." *Encyclopædia Britannica* at https://www.britannica.com/biography/Alexander-Graham-Bell (last accessed June 14, 2019).
Hornaday, William Temple. *The Extermination of the American Bison*. Washington, D.C.: Smithsonian Institution Press, c2002.
Hou, Jeffrey, Julie M. Johnson, and Laura J. Lawson. *Greening Cities, Growing Communities: Learning from Seattle's Urban Community Gardens.* Seattle, Washington: University of Washington Press, 2009.
House, Marguerita. "Lucretia Goes Buffalo Hunting," Buffalo Bill Center of the West, August 10, 2014, at http://centerofthewest.org/2014/08/10/lucretia-hunts-buffalo/ (last accessed October 2, 2015).
Huggins, David R., and John P. Reganold, "No-Till: the Quiet Revolution," *Scientific American* (June 30, 2008).
"Human Health Fact Sheet: Nitrate and Nitrite," Argonne National Laboratory, EVS (August 2005).
Huth, Hans. *Nature and the American: Three Centuries of Changing Attitudes*. Berkeley, California: University of California Press, 1957.
Illinois Department of Public Health, "Pyrethroid Insecticides" at http://www.idph.state.il.us/envhealth/factsheets/pyrethroid.htm (last accessed June 12, 2019).

"The Industrial Revolution (1750-1900)." *Encyclopædia Britannica* at https://www.britannica.com/technology/history-of-technology/The-Industrial-Revolution-1750-1900 (last accessed June 14, 2019).

International Union for Conservation of Nature. *Red List of Threatened Species* at http://www.iucnredlist.org/ (last accessed June 10, 2019).

International Union for Conservation of Nature. "Bison bison" in *Red List* at http://www.iucnredlist.org/details/2815/0 (last accessed June 10, 2019).

Iowa State University Extension. "Corn and Soybeans: Harvested Acreage and Yield per Acre" at http://www2.econ.iastate.edu/outreach/agriculture/periodicals/chartbook/Chartbook2/Tables/Table10.pdf (last accessed June 11, 2019).

Isenberg, Andrew C. *The Destruction of the Bison: An Environmental History, 1750-1920*. Cambridge, U.K.: Cambridge University Press, 2000.

Ives, Mike. "Recyclers Cringe as Southeast Asia Says It's Sick of the West's Trash." *New York Times*, June 7, 2019.

Jacobs, Julia. "Monsanto Ordered to Pay $80 Million in Roundup Cancer Case," *New York Times*, March 27, 2019.

James Madison and John Jay, *The Federalist*, ed. Benjamin Fletcher Wright. Cambridge, Massachusetts: The Belknap Press of Harvard University Press, 1961.

"Joint FAO/WHO Meeting on Pesticide Residues Summary Report," May 16, 2016, at https://www.who.int/foodsafety/jmprsummary2016.pdf (last accessed January 15, 2019).

"Judge Upholds Monsanto Verdict but Cuts Award to $78 Million," Associated Press, October 22, 2018.

"Jury Orders Monsanto to Pay Nearly $290 Million in Roundup Trial," *NBC News*, August 11, 2018, at https://www.nbcnews.com/news/us-news/jury-orders-monsanto-pay-290m-roundup-trial-n899811 (last accessed December 18, 2018).

Katz, Cheryl. "Piling Up: How China's Ban on Importing Waste Has Stalled Global Recycling." *Yale Environment 360*, March 7, 2019, at https://e360.yale.edu/features/piling-up-how-chinas-ban-on-importing-waste-has-stalled-global-recycling (last accessed June 14, 2019).

Kendall, M.G. "Losses of U.K. Merchant Ships in World War II," *Economica* (1948), 15:289.

Ketler, Alanna. "Sweden Runs Out of Garbage," November 26, 2014, at http://www.collective-evolution.com/2014/11/26/sweden-runs-out-of-garbage-only-1-ends-up-in-landfills/ (last accessed June 14, 2019).

Kolbert, Elizabeth. *The Sixth Extinction: An Unnatural History*. New York: Henry Holt and Company, 2014.

Krauss, Clifford. "U.S. Trying to Protect Sage Grouse Without Listing It as an Endangered Species." *New York Times*, September 22, 2015.

LaCapra, Veronique. "With 'Single-Stream' Recycling, Convenience Comes at a Cost." *National Public Radio (NPR)*, March 31, 2015 at http://www.npr.org/2015/03/31/396319000/with-single-stream-recycling-convenience-comes-at-a-cost (last accessed June 14, 2019).

Landau, Elizabeth. "Chick-fil-A To Serve Antibiotic-Free Chicken," *Cable News Network (CNN)*, February 12, 2014, at http://www.cnn.com/2014/02/11/health/chick-fil-a-chicken-antibiotics/ (last accessed June 12, 2019).

Langpap, Christian and JunJie Wu. "Voluntary Conservation of Endangered Species: When Does No Regulatory Assurance Mean No Conservation?" *Journal of Environmental Economics and Management*, June 2003, 47:435–457.

Langpap, Christian, and Joe Kerkvliet. "Endangered Species Conservation on Private Land: Assessing the Effectiveness of Habitat Conservation Plans." *Journal of Environmental Economics and Management*, July 2012, 64:14.

Langpap, Christian. "Conservation of Endangered Species: Can Incentives Work for Private Landowners?" *Ecological Economics* (2006), 57:558–572.

Larson, Lincoln R., *et al.* "Outdoor Time, Screen Time, and Connection with Nature: Troubling Trends Among Rural Youths?" *Environment and Behavior*, October 20, 2018, at https://journals.sagepub.com/doi/pdf/10.1177/0013916518806686 (last accessed April 18, 2019).

Laskow, Sarah. "Single-Stream Recycling Is Easier for Consumers, but Is It Better?" *The Atlantic*, September 18, 2014.

"Lausanne/Switzerland: Energy-from-Waste Plant" at http://www.hz-inova.com/cms/wp-content/uploads/2014/11/hzi_ref_lausanne-en.pdf (last accessed June 14, 2019).

Lawson, Laura J. *City Bountiful: A Century of Community Gardening in America*. Berkeley, California: University of California Press, 2005.

Lawton, Julia. "Is Conservation Agriculture Better than Organic Farming?" at http://science.howstuffworks.com/environmental/green-science/conservation-agriculture1.htm (last accessed June 12, 2019).

Leech, Brian James. *The City That Ate Itself: Butte, Montana and Its Expanding Berkeley Pit.* Reno, Nevada: University of Nevada Press, 2018.

Lemire, Beverly. "Shifting Currency: The Culture and Economy of the Second Hand Trade in England, c. 1600-1850," in *Old Clothes, New Looks*, edited by Alexandra Palmer and Hazel Clark. New York: Oxford International Publishers Ltd., 2005.

Leopold Center for Sustainable Agriculture at Iowa State University. "About the Leopold Center" at https://www.leopold.iastate.edu/about/leopold-center (last accessed April 22, 2019).

Leopold, Aldo. *A Sand County Almanac with Essays on Conservation from Round River*. New York: Ballantine Books, 1966.

Leopold, Aldo. "On a Monument to the Pigeon." *Birdwatching Daily* at https://www.birdwatchingdaily.com/blog/2014/09/01/monument-pigeon-aldo-leopold/ (last accessed June 10, 2019).

Levin, Sam. "The Man Who Beat Monsanto: 'They Have To Pay for Not Being Honest,'" *The Guardian*, September 26, 2018.

Lewis, Jack. "The Birth of EPA," *EPA Journal*, November 1985, at http://www2.epa.gov/aboutepa/birth-epa (last accessed October 12, 2015).

Lingam, Rama. "Mandatory Recycling Laws: Pros and Cons" at https://knoji.com/mandatory-recycling-laws-pros-and-cons (last accessed June 14, 2019).

Locke, John. *The Second Treatise of Government*. Edited by Thomas P. Peardon. Library of Liberal Arts Edition. Indianapolis, Indiana: Bobbs-Merrill Company, Inc., 1952.

Lombardo, Crystal. "Mandatory Recycling Pros and Cons," April 15, 2015, at http://www.visionlaunch.com/mandatory-recycling-pros-and-cons/ (last accessed January 16, 2019).

Louisiana Universities Marine Consortium. "Hypoxia in the Northern Gulf of Mexico" at http://www.gulfhypoxia.net/overview/ (last accessed December 16, 2016).

Madison, Erin. "First Peoples Named National Historic Landmark," *Great Falls Tribune* (Great Falls, Montana), August 4, 2015.

Madison, James. "Property." *National Gazette*, March 27, 1792, in *The Papers of James Madison*, Vol. 14. Edited by Robert A. Rutland *et al.* Charlottesville, Virginia: University of Virginia Press, 1983.

Mamiko, "A Primer on Recycling in Switzerland," January 28, 2012, at http://www.newlyswissed.com/a-primer-on-recycling-in-switzerland-2/ (last accessed June 14, 2019).

McDonald's Global Policy on Antibiotic Use in Food Animals, at https://www.mcdonalds.com/dam/AboutMcDonalds/Sustainability/Sustainability%20Library/antibiotics_policy.pdf (last accessed June 12, 2019).

McDonald's Global Vision for Antimicrobial Stewardship in Food Animals at http://worldvet.org/uploads/news/docs/antimicrobial_stewardship_vision.pdf (last accessed June 12, 2019).

McEwen, Scott A. "Antibiotic Use in Animal Agriculture: What We Have Learned and Where Are We Going?" *Animal Biotechnology* (2006) 17: 243.

McNeil, Donald G., Jr. "Upstate Waste Site May Endanger Lives." *New York Times*, August 2 1978.

McVeigh, Karen. "Scientists: Overuse of Antibiotics in Animal Agriculture Endangers Humans," *Guardian*, September 19, 2012.

"Merchant Navy Losses" at http://www.worldnavalships.com/merchant_navy_losses.htm (last accessed June 13, 2019).

Merrild, Hanna, Anna W. Larsen, and Thomas H. Christensen. "Assessing Recycling Versus Incineration of Key Materials in Municipal waste: The Importance of Efficient Energy Recovery and Transport Distances." *Waste Management*, May 2012.

Metzger, Bruce M., and Roland E. Murphy, editors. *The New Oxford Annotated Bible*. Oxford, U.K.: Oxford University Press, 1991.

Mink, Pamela J., Jack S. Mandel, Bonnielin K Sceuman, and Jessica I. Lundin, "Epidemiologic Studies of Glyphosate and Cancer: A Review," *Regulatory Toxicology and Pharmacology* (2012), 63:440–452 at http://www.sciencedirect.com/science/article/pii/S0273230012000943 (last accessed September 10, 2016).

Minnesota Department of Agriculture. "Guide for Minnesota Farmers and Ranchers Living in Wolf Territory" at http://www.mda.state.mn.us/grants/disaster/wolf.aspx (last accessed September 3, 2016).

Montana Bureau of Mines and Geology. "Environmental Assessment—Berkeley Pit" at https://www.mbmg.mtech.edu/env/env-berkeley.asp (last accessed December 9, 2016).

Montana Land Reliance. "Conservation Easements" at http://mtlandreliance.org/conservation-easements/ (last accessed April 22, 2019).

"Montana Livestock Loss Board" at http://liv.mt.gov/llb/default.mcpx (last accessed October 21, 2015).

Montana State Parks, "Madison Buffalo Jump State Park" at http://stateparks.mt.gov/madison-buffalo-jump/ (last accessed June 10, 2019).

Montana State Parks. "First Peoples Buffalo Jump State Park" at http://stateparks.mt.gov/first-peoples-buffalo-jump/ (last accessed June 10, 2019).

Murphy, James E. *Half Interest in a Silver Dollar: The Saga of Charles E. Conrad*. Missoula, Montana: Mountain Press Publishing Company, 1983.

National Conservation Easement Data Base at https://www.conservationeasement.us/ (last accessed April 22, 2019).

National Oceanic and Atmospheric Administration (NOAA), "What Is Eutrophication?" at https://oceanservice.noaa.gov/facts/eutrophication.html (last accessed June 11, 2019).

National Pesticide Information Center (NPIC). "Glyphosate Technical Fact Sheet" at http://npic.orst.edu/ingred/glyphosate.html (last accessed June 12, 2019).

Nature Conservancy. "Conservation Easements" at http://www.nature.org/about-us/private-lands-conservation/conservation-easements/all-about-conservation-easements.xml (last accessed April 22, 2019).

Nelson, Donald M. *Arsenal of Democracy: The Story of American War Production*. New York: Harcourt, Brace and Co., 1946.

NOAA Ocean Service Education. "Nutrient Pollution—Eutrophication," at http://oceanservice.noaa.gov/education/kits/estuaries/media/supp_estuar09b_eutro.html (last accessed April 22, 2019).

Northern Virginia Soil and Water Conservation District. "No-Till Agriculture: Good for the Soil and the Bottom Line." *Conservation Currents*, September 2005, at http://www.fairfaxcounty.gov/nvswcd/newsletter/notill.htm (last accessed June 12, 2019).

Nowatzki, John. "GIS Application of Potential Contamination of Ground water from Agricultural Pesticides" at http://proceedings.esri.com/library/userconf/proc00/professional/papers/pap198/p198.htm (last accessed June 11, 2019).

"Oregon Department of Agriculture Will Compensate Ranchers for Livestock Killed by Wolves" at http://www.oregonlive.com/environment/index.ssf/2011/12/oregon_begins_compensation_pro.html (last accessed June 10, 2019).

Oregon State University Extension Service. "What Do the Numbers of Fertilizer Packages Mean?" at http://extension.oregonstate.edu/gardening/node/679 (last accessed June 11, 2019).

Organic Foods Production Act of 1990 at https://www.ams.usda.gov/sites/default/files/media/NOPOFPA.pdf (last accessed June 12, 2019).

Organic Trade Association. "U.S. Organic Sales Break through $50 Billion Mark in 2018" at https://ota.com/news/press-releases/20699 (last accessed June 16, 2019).

Original Journals of the Lewis and Clark Expedition 1804-1806, Vol. II, Part 1. New York: Dodd, Meade & Company, 1904.

Paddock, Catharine. "FDA to Phase out Non-Therapeutic Use of Antibiotics in Food Animals" *Medical News Today*, December 12, 2013 at http://www.medicalnewstoday.com/articles/270010.php (last accessed June 11, 2019).

"Passenger Pigeon: Demography and Populations." *Birds of North America Online* at http://bna.birds.cornell.edu/bna/species/611/articles/demography?searchterm=Passenger%20Pigeon.

"Paul IV" at https://www.ccjr.us/dialogika-resources/primary-texts-from-the-history-of-the-relationship/paul-iv (last accessed June 13, 2019).

Paul VI, *Populorum Progressio,* March 26, 1967, at http://w2.vatican.va/content/paul-vi/en/encyclicals/documents/hf_p-vi_enc_26031967_populorum.html (last accessed April 30, 2019).

Paul, Bill, Ed Bangs, Carter Niemeyer, and Elizabeth Harper. "Wolf Depredation," January 2014 at http://www.defenders.org/press-release/defenders-shifts-focus-wolf-coexistence-partnerships (last accessed June 10, 2019).

Peace, W.S. "Sharps Buffalo Rifle." *Handbook of Texas Online* at https://tshaonline.org/handbook/online/articles/lns02 (last accessed June 10, 2019).

Perfect Rubber Mulch at https://perfectrubbermulch.com/ (last accessed June 14, 2019).

Pestizid Aktions Netzwerk (PAN), "Phasing in Alternatives to DDT" at http://www.pan-germany.org/download/ddt/phasing_in_alternatives_to_ddt.pdf (last accessed June 12, 2019).

Pew Commission on Industrial Farm Animal Production, *Putting Meat on the Table: Industrial Farm Animal Production in America*, April 29, 2008.

Pierce, Ben. "Places We Love: First Peoples Buffalo Jump State Park," *Bozeman Daily Chronicle* (Bozeman, Montana), April 16, 2015.

Plumer, Brad. "No-Till Farming Is on the Rise. That's Actually a Big Deal." *Washington Post* (Washington, D.C.), November 9, 2013.

Ponting, Clive. *A Green History of the World*. London, U.K.: Penguin Books, 1992.

Potter, Ralph. *War and Moral Discourse.* Richmond, Virginia: John Knox Press, 1969.

Quote Investigator at http://quoteinvestigator.com/2011/10/20/last-tree-cut/ (last accessed April 22, 2019).

Reuters. "Monsanto Sues California over Herbicide Classification," *New York Times*, January 21, 2016.

Reuters. "W.H.O. Report Links Ingredient in Roundup to Cancer." *New York Times*, March 20 2015.

Rolston, Holmes III. *Environmental Ethics: Duties to and Values in the Natural World.* Philadelphia, Pennsylvania: Temple University Press, 1988.

"Rome" in *Encyclopaedia Britannica Academic* at http://academic.eb.com/EBchecked/topic/508807/Rome/23920/Slow-decline-of-the-late-empire (last accessed June 13, 2019).

Samsel, Anthony, and Stephanie Seneff. "Glyphosate's Suppression of Cytochrome P450 Enzymes and Amino Acid Biosynthesis by the Gut Microbe: Pathways to Modern Diseases," *Entropy* [2003], 15:1416.

Sandoz, Mari. *The Buffalo Hunters: The Story of the Hide Men.* New York: Hastings House Publishers, 1954.

Santa Barbara County Resource Recovery & Waste Management Division. "Why Recycle?" at http://www.lessismore.org/materials/28-why-recycle (last accessed June 14, 2019).

Siegan, Bernard. *Property and Freedom: The Constitution, the Courts, and Land-Use Regulation.* New Brunswick, Canada: Transaction Publishers, 1997.

Silberner, Joanne. "WHO Backs Use of DDT Against Malaria," National Public Radio, 15 September 2006, at http://www.npr.org/templates/story/story.php?storyId=6083944 (last accessed June 12, 2019).

Souder, William. "100 Years After Her Death, Martha, the Last Passenger Pigeon Still Resonates." *Smithsonian Magazine*, October 2014.

Stanley, Anne. "Sustainable America: West Coast Cities Leading the Way with Recycling." *American City & County*, March 4, 2015, at http://web.a.ebscohost.com/ehost/detail/detail?sid=cdf5ba5a-d32d-450a-9133-b68495afb908%40sessionmgr4004&vid=11&hid=4207&bdata=JnNpdGU9ZWhvc3QtbGl2ZQ%3d%3d#AN=101387043&db=a9h (last accessed January 16, 2019).

Stephen R. Kellert *et al. The Nature of Americans: Disconnection and Recommendations for Reconnection.* Nature of Americans.org, April 2017 at https://natureofamericans.org/sites/

default/files/reports/Nature-of-Americans_National_Report_1.3_4-26-17.pdf (last accessed April 18, 2019).
Stockholm Convention on Persistent Organic Pollutants at http://chm.pops.int/Convention/ConventionText/tabid/2232/Default.aspx (last accessed June 12, 2019).
Strasser, Susan. *Waste and Want: A Social History of Trash.* New York: Henry Holt and Company, 1999.
Supreme Court of the United States. *Babbitt v. Sweet Home Chapter of Communities for a Great Oregon*, 515 U.S. 687 (1995).
Supreme Court of the United States. *Lucas v. South Carolina Coastal Council*, 505 U.S. 1103 (1992).
Supreme Court of the United States. *Michigan v. Environmental Protection Agency*, 576 U.S. (2015).
Supreme Court of the United States. *Miller v. Schoene*, 276 U.S. 272 (1928).
Supreme Court of the United States. *Penn Central Transportation Co. v. New York City*, 438 U.S. 104 (1978).
Supreme Court of the United States. *Pennsylvania Coal v. Mahon*, 260 U.S. 393 (1922).
Supreme Court of the United States. *Tennessee Valley Authority v. Hill et al.*, 437 U.S. 153 (1978).
The Act of Dedication—March 1, 1872 (U.S. Code, Title 16, Chapter 1, Subchapter V) at http://www.yellowstone-online.com/history/yhfour.html (last accessed June 10, 2019).
Thongprakaisang, Siriporn *et al.*, "Glyphosate Induces Human Breast Cancer Cells Growth via Estrogen Receptors, *Food and Chemical Toxicology* (2013), 59:129–136
Thoreau, Henry David. *Walden and Other Writings of Henry David Thoreau*, edited by Brooks Atkinson. New York: The Modern Library, 1950.
"Timeline of the Great Depression" at http://www.hyperhistory.com/online_n2/connections_n2/great_depression.html (last accessed June 13, 2019).
Townsend, John Kirk. "Across the Rockies to the Columbia" at http://user.xmission.com/~drudy/mtman/html/townsend/chap3.html (last accessed June 10, 2019).
United States Court of Federal Claims. *Good v. United States*, 39 Fed.Cl. 81 (1997).
United States Census Bureau, *Statistical Abstract of the United States: 2012*, Table 861 (551).
United States Department of Agriculture National Institute of Food and Agriculture, "Legal Definition of Sustainable Agriculture" at https://nifa.usda.gov/program/sustainable-agriculture-program (last accessed June 12, 2019).
United States Department of Agriculture Natural Resources Conservation Service (NRCS). "Sage Grouse Initiative" at http://www.sagegrouseinitiative.com/about/ (last accessed June 10, 2019).
United States Department of Agriculture Natural Resources Conservation Service (NRCS). *Sage Grouse Initiative: Success on the Range* at http://www.sagegrouseinitiative.com/wp-content/uploads/2015/07/SGI-Success-on-the-Range-FINAL-LOW-RES-FULL-SPREADS-WITH-HYPERLINKS-FOR-WEB-073015.pdf (last accessed June 10, 2019).
United States Department of Agriculture, "Recent trends in GE Adoption" at https://www.ers.usda.gov/data-products/adoption-of-genetically-engineered-crops-in-the-us/recent-trends-in-ge-adoption.aspx (last accessed June 12, 2019).
United States Department of Agriculture. "Introduction to Organic Practices" at https://www.ams.usda.gov/sites/default/files/media/Organic%20Practices%20Factsheet.pdf (last accessed June 12, 2019).
United States Department of Agriculture. *Summary Report: 2015 National Resources Inventory*, USDA Natural Resources Conservation Service, Washington, D.C., and Center for Survey Statistics and Methodology, Iowa State University, Ames, Iowa, (September 2018) at https://www.nrcs.usda.gov/Internet/FSE_DOCUMENTS/nrcseprd1422028.pdf (last accessed June 11, 2019)
United States Department of Agriculture. "Organic Regulations" at https://www.ams.usda.gov/rules-regulations/organic (last accessed June 12, 2019).
United States Department of Energy Alternate Fuels Data Center. "U.S. Corn Productivity and Land Use" at http://www.afdc.energy.gov/data (last accessed June 11, 2019).

United States Environmental Protection Agency. "Recycling, Energy Conservation, and Community Beautification" at http://www.epa.gov/region03/beyondtranslation/2013BTF/SessionB-Beautification/MichelleFeldman.pdf (last accessed September 1, 2015).

United States Environmental Protection Agency. "Draft Human Health and Ecological Risk Assessments for Glyphosate" at https://www.epa.gov/ingredients-used-pesticide-products/draft-human-health-and-ecological-risk-assessments-glyphosate (last accessed June 8, 2018).

United States Environmental Protection Agency. "EPA Releases Draft Risk Assessments for Glyphosate" at https://www.epa.gov/pesticides/epa-releases-draft-risk-assessments-glyphosate (last accessed June 8, 2018).

United States Environmental Protection Agency. "Basic Information about Nitrate in Drinking Water" at http://water.epa.gov/drink/contaminants/basicinformation/nitrate.cfm (last accessed June 11, 2019).

United States Environmental Protection Agency. "Integrated Pest Management" at http://www.epa.gov/agriculture/tipm.html (last accessed January 29, 2013).

United States Fish and Wildlife Service. "A History of the Endangered Species Act of 1973" at http://www.fws.gov/endangered/laws-policies/esa-history.html (last accessed June 10, 2019).

United States Fish and Wildlife Service. "Endangered Species Act Timeline" at http://www.fws.gov/endangered/laws-policies/timeline.html (last accessed June 10, 2019).

United States Fish and Wildlife Service. "National Bison Range" at http://www.fws.gov/refuge/National_Bison_Range/about.html (last accessed June 19, 2019).

United States Fish and Wildlife Service. "Time Line of the American Bison" at http://www.fws.gov/bisonrange/timeline.htm (last accessed June 10, 2019).

United States Fish and Wildlife Service. "Time Line of the American Bison" at http://www.fws.gov/bisonrange/timeline.htm (last accessed April 22, 2019).

United States Food and Drug Administration, "Drugs Transitioning from Over-the-Counter (OTC) to Veterinary Feed Directive (VFD) Status" at https://www.fda.gov/downloads/AnimalVeterinary/DevelopmentApprovalProcess/UCM482258.pdf (last accessed June 12, 2019).

United States Food and Drug Administration. "About NARMS" at https://www.fda.gov/AnimalVeterinary/SafetyHealth/AntimicrobialResistance/NationalAntimicrobialResistanceMonitoringSystem/ucm059089.htm (last accessed June 12, 2019).

United States Food and Drug Administration. "New Animal Drugs; Cephalosporin Drugs; Extralabel Animal Drug Use; Order of Prohibition," *Federal Register* (January 6, 2012), 77, 4: 738.

United States Food and Drug Administration. "Veterinary Feed Directive (VFC)" at https://www.fda.gov/AnimalVeterinary/DevelopmentApprovalProcess/ucm071807.htm (last accessed June 12, 2019).

United States Food and Drug Administration. "Fact Sheet: Veterinary Feed Directive Final Rule and Next Steps" at https://www.fda.gov/AnimalVeterinary/DevelopmentApprovalProcess/ucm449019.htm (last accessed June 8, 2018).

United States Forest Service History. "1979: Snail Darter Exemption Case" at http://www.foresthistory.org/ASPNET/Policy/northern_spotted_owl/1979owl.snaildarter.aspx (last accessed October 12, 2015).

United States Geological Survey. "Pesticides in the Nation's Streams and Ground Water, 1992-2001—A Summary," at http://pubs.usgs.gov/fs/2006/3028/pdf/fs2006-3028.pdf (last accessed June 11, 2019).

United States National Park Service. "Bison Management—Yellowstone National Park" at http://www.nps.gov/yell/learn/nature/bisonmgnt.htm (last accessed October 1, 2015).

United States National Park Service. "Visitation Numbers," March 6, 2019, at https://www.nps.gov/aboutus/visitation-numbers.htm (last accessed April 18, 2019).

University of Illinois Extension. "Integrated Pest Management" at http://www.extension.org/pages/17849/integrated-pest-management (last accessed January 29, 2013).

University of Montana. "Grizzly Bear Ecosystems" at http://www.cfc.umt.edu/grizzlybearrecovery/grizzly-bears/ecosystems.php (last accessed June 10, 2019).

"USDA, Montana Enter Public-Private Agreement to Enhance Montana's Sage Grouse Conservation Efforts" at http://www.nrcs.usda.gov/wps/portal/nrcs/detail/national/newsroom/releases/?cid=NRCSEPRD377619 (last accessed June 10, 2019).

Weber, Max. *The Protestant Ethic and the Spirit of Capitalism*. Translated by Talcott Parsons. New York: Charles Scribner's Sons, 1958.

Western Association of Fish and Wildlife Agencies. "WAFWA Report Documents Greater Sage-Grouse Population Rebound," August 17, 2015, at http://www.wafwa.org/news/e_1606/News/2015/8/_WAFWA_Report_Documents_Greater_Sage-Grouse_Population_Rebound.htm (last accessed June 10, 2019).

"Why is Recycling Not Mandatory in All U.S. Cities?" *Earthtalk* at http://environment.about.com/od/recycling/a/must_recycle.htm (last accessed January 16, 2019).

Wilcove, David S., Michael J. Bean, Robert Bonnie, and Margaret McMillan. *Rebuilding the Ark: Toward a More Effective Endangered Species Act for Private Land*, December 5, 1996, at http://omnilearn.net/esacourse/pdfs/Rebuilding_the_Ark.pdf (last accessed June 10, 2019).

Wildlife Conservation Society. "The American Bison Society" at http://www.wcs.org/saving-wildlife/hoofed-mammals/bison/the-american-bison-society.aspx (last accessed June 10, 2019).

Williams, Gary M., Robert Kroes, and Ian C. Munroe, "Safety Evaluation and Risk Assessment of Herbicide Roundup and Its Active Ingredient, Glyphosate, for Humans," *Regulatory Toxicology and Pharmacology* (2000), 31:117–165 at http://www.msal.gob.ar/agroquimicos/pdf/Williams-et-al-2000.pdf (last accessed September 10, 2016).

Wooster, Robert. *The Military and United States Indian Policy 1865-1903*. New Haven, CT: Yale University Press, 1988.

World Health Organization (WHO). "Antimicrobial Resistance" at http://www.who.int/mediacentre/factsheets/fs194/en/# (last accessed June 11, 2019).

Xiao Lim. "Is Glyphosate, Used with Some GM crops, Dangerously Toxic to Humans?" *Genetic Literacy Project*, April 30, 2014, at https://www.geneticliteracyproject.org/2014/04/30/is-glyphosate-used-with-some-gm-crops-dangerously-toxic-to-humans/ (last accessed June 12, 2019).

Yeoman, Barry. "Why the Passenger Pigeon Went Extinct." Audubon Magazine, May–June 2014.

Zaveri, Mihir. "Monsanto Weedkiller Roundup Was 'Substantial Factor' in Causing Man's Cancer, Jury Says," *New York Times*, March 19, 2019.

Zimring, Carl A. *Cash for Your Trash: Scrap Recycling in America*. New Brunswick, New Jersey: Rutgers University Press, 2005.

Index

About the Author

Daniel E. Lee is the Marian Taft Cannon Professor in the Humanities at Augustana College, Rock Island, Illinois, where he has been a member of the faculty since 1974. Along with his teaching responsibilities, he serves as Director of the Augustana Center for the Study of Ethics. He is the author of numerous articles and ten books, including *Human Rights and the Ethics of Globalization*, which he coauthored with his daughter, Elizabeth J. Lee. *Human Rights and the Ethics of Globalization* was a Choice Outstanding Academic Title 2011 winner. He has served as a guest professor in the College of Philosophy at Shanghai Normal University and as a visiting professor at Macquarie University in Sydney, Australia. Born in a mountain valley in northwestern Montana, he attended a two-room country school with a wood-burning stove. A summa cum laude graduate of Concordia College, Moorhead, Minnesota, he received the MA, MPhil, and PhD degrees from Yale University.

www.ingramcontent.com/pod-product-compliance
Ingram Content Group UK Ltd.
Pitfield, Milton Keynes, MK11 3LW, UK
UKHW041122071026
15587UKWH00009B/97

9 781498 599313